TELECOMMUNICATIONS SWITCHING PRINCIPLES

Telecommunications Switching Principles

M. T. HILLS

The MIT Press
Cambridge, Massachusetts, and London, England

First MIT Press edition, 1979

First published in England in 1979 by
George Allen & Unwin Ltd.

© 1979 M. T. Hills
All rights reserved. No part of this book may be reproduced in any form or by any means, electronic or mechanical, including photocopying, recording, or by any information storage and retrieval system, without permission in writing from the publisher.

Printed in Great Britain

Library of Congress catalog card number: 87-61815
ISBN 0-262-08092-3

*Dedicated to
A. Strowger
the originator of the world's
first and largest real-time multi-processor
control system*

Preface

Telecommunications networks for public and private use are possibly among the largest interconnected systems in the world. The world-wide public telephone service is provided by a network interconnecting more than 350 million telephones. A telecommunications network consists of terminals, transmission links and switching centres. The principles underlying the design of the transmission and terminal components are well established and have a coherence. In contrast there appears to be a multitude of different approaches to the design of the switching centre component. One object of this book is to develop a coherent set of principles for the design of these centres.

Several practical examples are described of different systems around the world to demonstrate the application of the principles and to give the reader a background knowledge of the state of art. Most of these examples are taken from the public telephone network as it is in this network the principles were evolved. However, the principles are shown to apply equally to other types of switching systems such as data networks or private telephone systems. The original intention was to include a wide range of system descriptions but reasons of economy dictated their omission. The choice of systems actually included was random and omissions include many of the major systems currently available.

The majority of switching equipment in public use today is still built from electro-mechanical components such as relays. Much of this technology will be unfamiliar to recently qualified engineers and some of the book is devoted to describing its application.

The use of computers and, more recently, micro-processors is bringing a revolution in the design of switching systems, and a significant proportion of the book is devoted to such systems. However, it is suggested (in Chapter 13) that some of the earlier design of computer-controlled switching centres could have been improved if they had more fully adopted the design principles developed for the electro-mechanical systems.

The book should be of value to an engineer with an electronic engineering, computer systems or conventional switching background who has either to design or to operate a switching system or who needs to use one. It could form the basis of a final year course in telecommunication engineering, teleprocessing or real-time computer systems. Apart from some mathematics associated with the traffic and queueing theory chapter, the only prerequisite knowledge is that of basic electronics and in certain sections a background knowledge of computers and programming.

PREFACE

The book is based on courses given at postgraduate level in the University of Essex M.Sc. in Telecommunications Systems, to numerous classes of new entrants to the British Post Office and to many audiences around the world.

The book in its current form has been made possible by the interaction with numerous colleagues within the Electrical Engineering Department, University of Essex and in the British Post Office, GEC, Plessey, Standard Telephone and Cables, Pye-TMC and Thorn-Ericsson in the U.K. Abroad, the strongest influence has been from the Nippon Telegraph and Telephone Public Corporation (NTT) where I spent a sabbatical leave in their Electrical Communications Laboratories in the Tokyo area. I am particularly grateful to Malcolm Hamer who has painstakingly read several drafts of the book and in many detailed discussions assisted me in removing much of the obscurity in the earlier drafts.

I should like to express my gratitude to the British Post Office whose financial support of the Telecommunications group within the Department of Electrical Engineering Science has made possible the environment in which the concepts described in the book have been developed and tested. My hope is that some of the ideas discussed in the book will go towards improving the design of telephone switching systems.

Finally, I must thank my publisher Roger Jones for his apparently groundless faith that I would actually get the book finished albeit five years later than originally planned. Without his continued encouragement even this version would not have appeared.

Contents

		page
1	*Introduction to switching systems*	
	1.1 Purpose of the book	1
	1.2 Types of switching systems	3
	1.3 Centralised switching systems	4
	1.4 Basic switching centre model	12
	1.5 Resource sharing	18
	1.6 Introduction to traffic and queueing theory	21
2	*Signalling and switching techniques*	
	2.1 Telephone systems	27
	2.2 Data and message switching systems	32
	2.3 Switching elements	36
3	*The design of economic switching centres*	
	3.1 Introduction	54
	3.2 Functional subdivision	55
	3.3 Common switch networks and their control	62
	3.4 Control of switching systems	64
	3.5 Alternative organisations of switching networks	72
	3.6 Time-shared decision making	76
	3.7 Stored program control	83
4	*Traffic theory*	
	4.1 Basic equations	86
	4.2 Queueing systems	92
	4.3 Comparison of the various distributions	93
	4.4 Applications of traffic theory	97
	4.5 Multi-stage switch networks – conditional selection	105
	4.6 Simulation techniques	108
	4.7 Traffic measurement and traffic prediction	111
5	*Telephone network organisation*	
	5.1 Network planning	113
	5.2 Routing plans	114
	5.3 Numbering plans	129
	5.4 Register control of networks	132
	5.5 Use of step-by-step equipment	134
	5.6 Charging	137
	5.7 Network management	138

xii CONTENTS

6 *Practical signalling systems*
   6.1 Types of signal for telephone systems   143
   6.2 Signalling techniques   147

7 *Design of switching networks*
   7.1 Basic multi-stage networks   160
   7.2 Use of mixing stages   167
   7.3 Network and channel graphs   170
   7.4 Networks with concentration   175
   7.5 Lee's simulation technique for evaluating blocking probabilities   179

8 *Control unit design*
   8.1 Role of control units   181
   8.2 The arbiter   182
   8.3 State transition diagram for a called control unit   184
   8.4 Signalling between control units   193
   8.5 Signalling techniques   195
   8.6 Design of a control unit for two-way traffic   198

9 *Some circuit techniques*
   9.1 Basic elements   205
   9.2 Electronic components   213
   9.3 Realisation of arbiters   215
   9.4 Third wire control   218
   9.5 Junctors   222
   9.6 Switch matrix control   226

10 *Practical examples of switching systems*
   10.1 Introduction   232
   10.2 Step-by-step   232
   10.3 Crossbar systems   239
   10.4 Reed relay systems   250

11 *Computer controlled switching systems*
   11.1 Software organisation of computer controlled centres   258
   11.2 No. 1 ESS   267
   11.3 Japanese D-10   277
   11.4 Metaconta systems   284

12 *Digital switching systems*
   12.1 Time-division switches   287
   12.2 Relative merits of digital and analogue switching   293
   12.3 Practical application of digital telephone system – Bell No. 4 ESS   295

CONTENTS xiii

13	The future – a personal opinion	
	13.1 A critique of switching system design	296
	13.2 Design principles for switching centre design	299
	13.3 In summary	301

Appendices

A.	Optimal size of switching machine	303
B.	Derivation of some basic traffic theory results	306
C.	Lower limit on number of crosspoints needed in multi-stage networks	311
D.	Simplified proof of Takagi optimum channel graph theorem	314
E.	Estimation of the traffic capacity of two-stage group selectors	322

Tables

1.1	Names of different level switching centres in telephone networks	10
1.2	Some typical availability objectives for public telephone systems	12
2.1	Example of telex signalling	33
2.2	Example of F31 telegram message format	34
4.1	Some results for a two-server system	94
4.2	Results for two-server system with infinite queue	94
4.3	Relationship between offered traffic level and server efficiency	99
4.4	Blocking probabilities of networks shown in Figure 4.12	108
5.1	Comparison of mesh and star connections	116
5.2	Calculation of alternative routing strategy	121
6.1	R1 signalling system (N. American system)	152
6.2	R2 signalling system	153
7.1	Blocking probabilities for simple multi-stage networks	167
7.2	Comparison of networks of Figure 7.12	179
11.1	Causes of down time in No. 1 ESS	275
D.1	Values of $F(i,j)$ corresponding to different states of the two E links	317
D.2	Values of B_a and B_b for $a = \frac{1}{2}$	319

1
Introduction to switching systems

1.1 Purpose of the book

The world-wide public telephone system is remarkable by virtue of the wide variety of equipment used in its construction from mechanical devices installed in the late 1920s to the most modern miniaturised digital circuitry. It all has to work together to provide an economic and reliable service.

Although it is the most visible switching system, the telephone system is not the only one. Other examples of switching systems are the public telex network allowing dial-up communication between teleprinters, and a range of national and international networks for company, bank and military users. Of increasing importance during the 1980s will be the growth of data switching which many people predict will overtake voice communication (in volume) by the end of the century. There are other systems which at first sight may not be regarded as switching systems but whose design, as this book shows, is based on the same principles. These include telemetry systems, on-line access to central computers from widely dispersed devices such as visual display terminals, card readers and so on.

The purpose of a telecommunications switching system is to provide the means to pass information from any terminal device to any other terminal device selected by the originator. Three components are necessary for such systems:

terminals which are either input or output transducers. They convert the information into an electrical signal at the transmitting end and convert the electrical signal back into a usable form at the receiving end. A further function of a terminal is to generate and transmit control signals to indicate the required destination of the information signal.
transmission links to convey the information and control signals between the terminals and switching centres.
switching centres to receive the control signals and to forward or connect the information signals.

Transmission links are covered in detail in a companion volume [1] and receive only scant mention in this book. Terminals are covered only from the point of view of their capabilities to generate and receive the control signals. The book deals with the design of the individual switching centres and their incorporation in switching networks.

The organisation of this book has two aims:

(1) To show that there is a unified set of principles behind the wide range of superficially different switching centre designs, and to show how these principles may be applied to switching centre design using modern technology.
(2) To give a description of the implementation of the design principles in some of the switching centre types in use today.

This chapter introduces the basic systems concepts and the objectives of a system design. Chapter 2 describes the basic signalling and switching technique used in voice and data switching centres. The design of economic switching centres is covered in Chapter 3. Economic design involves resource sharing and resource sharing implies that there is a probability that a resource (such as a switch or a transmission link) will not be available at the instant it is required. The design of a switching centre therefore involves determining the number of resources required to achieve a particular probability of no resource being available (and possibly a particular length of wait for a resource to become available). This is the subject of traffic theory which is covered in Chapter 4.

Switching centres are organised in networks and these are discussed in Chapter 5. The practical means by which control signals are passed from centre to centre are the subject of Chapter 6.

The next two chapters are more theoretical and in Chapter 7 a coherent approach is given to the design of switch networks. Of particular importance in this chapter is a discussion of the work of Takagi on optimum channel graphs. It is thought that this is the first time that this treatment has been covered in a text book.

Chapter 8 is a theoretical approach to the design of control systems and attempts to show a unified approach to the design of electronic and computer controlled systems and the relationship to signalling systems.

After the theoretical chapters, the remainder of the book deals with practical aspects of telephone switching systems. Chapter 9 describes a wide range of the practical techniques used within electro-mechanical and electronic systems. A selection of electro-mechanical and electronic systems are described in Chapter 10 and the examples are carefully chosen to demonstrate the application of particular principles. Computer controlled systems have Chapter 11 to themselves and some of this chapter requires a basic knowledge of computers.

The long-term future for the technology of switching systems is almost certainly going to be digital and the basic differences to other systems together with an example are discussed in Chapter 12.

Finally, the author has allowed himself some licence and the last chapter is in the form of an editorial, which attempts to predict the future direction of system architectures. This chapter attempts to show that the centralised control systems origins come from the 1960s and are not the way today's systems should be designed. The centralised systems appear to have been designed as an exercise in programming rather than as the design of a reliable telephone switching system using principles described in this book. Time will show who was right.

1.2 Types of switching systems

A system similar to (but smaller than) the telephone network is the public telex network (*tele*graph *ex*change) which provides world-wide direct interconnection of teleprinters. Both the telephone and telex networks are examples of what are called *circuit switching* since they set up a circuit between two terminals which then interchange information directly.

Another class of system which is more familiar to the business or military user is that of *message switching*. The terminals of message switched systems are usually teleprinters, but unlike the telex network they are not interconnected directly. Instead, when a terminal user types a message destined for some other terminal, the system stores the message and delivers it to the required terminal at some later time. The reason for the delay is that the system is designed to maximise the utilisation of transmission links by queueing messages awaiting the use of a link. In order to set up a direct connection over many links connected end-to-end it is necessary for each link to be simultaneously free. As will be seen later, this implies that the average utilisation of the links must be low if the probability of a direct connection being available on demand is to be high enough to satisfy most users. However, in a message switched system, where messages are queued for each link, a much higher link utilisation is achieved. Another name for this type of system is *store and forward* switching.

The advent of real-time computer systems for airline reservation, banking systems and remote data processing in general has been based on the use of telecommunications networks to carry data between computer-type terminal devices and large real-time computers. Such applications may be served by a purpose-built circuit switched network or by another system such as the public telephone network in conjunction with special signal processing techniques.

More recently, general purpose *packet switching systems* have been developed which take the data from a terminal or a computer and transmit it as short packets of information to the required destination. Such systems are midway between the two extremes of circuit switching and store and forward. The interchange of packets may be made so rapid that a terminal appears to provide a 'conversational' connection while at the same time high transmission link utilisations are obtained through queueing.

1.3 Centralised switching systems

A simple way of structuring a switched network is to arrange that each terminal has a direct transmission link to every other terminal, as shown in Figure 1.1a. Each terminal needs a switch to connect it to the required link and a switch to make connection to a link in order to receive an incoming call. For N terminals this arrangement needs a total of $\tfrac{1}{2}N(N-1)$ links.

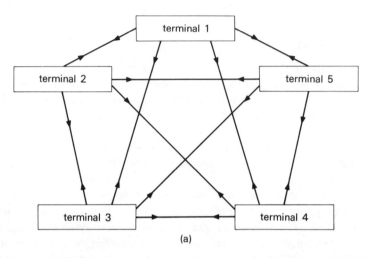

Figure 1.1 (a) Full interconnection for five terminals. Some form of switching is needed within each terminal to connect it to the appropriate link. (b) Use of one channel per terminal. Each terminal is permanently connected to one channel and all other terminals may access a particular terminal by operating a switch which connects it to the appropriate channel. (c) As in (b) but with less than N channels.

An alternative approach which needs only N links is to provide one link per terminal and to arrange that all other terminals have access to it as shown in Figure 1.1b. This simplifies the terminal equipment because it removes the need to connect a terminal to a link for an incoming call. This is a practical arrangement for systems such as house telephones or intercoms where there is a relatively small number of terminals close together. For instance, it is possible to provide an eleven-terminal system with each terminal having ten buttons and eleven pairs of wires going around to each instrument. However, when the number of terminals increases, or their geographical separation increases, the cost of cabling makes this arrangement uneconomic. Another example of the technique illustrated in Figure 1.1b is a radio telephone network in which each terminal is given its own frequency. An originator can set-up a call by tuning his transmitter to the frequency of the called terminal.

CENTRALISED SWITCHING SYSTEMS

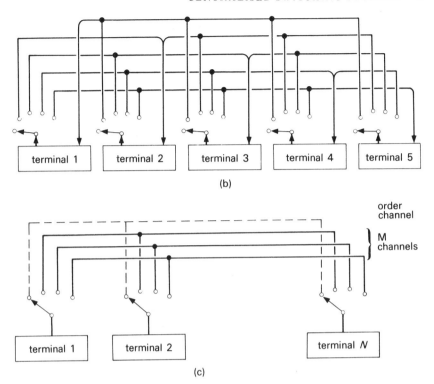

(b)

(c)

It is possible to use an arrangement similar to that just described but with fewer than N links, as shown in Figure 1.1c. With this arrangement it is necessary for the calling terminal to select a free link and for the called terminal to be connected to the same link. One technique used extensively for radio telephones involves a common channel to which all free terminals are connected [2]. A calling signal (which may be coded or verbal) is sent along this link and is received by the terminals. The calling signal informs the called terminal (or its user) of an incoming call and indicates the link to which the terminal should be switched in order to receive the call. The recognition of the calling signal and the switching operations may be performed automatically in a system using coded signals. In simpler systems the users may be required to listen out for, and act upon, verbal calling signals.

At the present time this technique of 'distributed' switching is applied only to small telephone systems and to some radio telephone networks. One of the possibilities of the future is what is called a 'digital ring main' in which a high speed binary digital link is connected to a large number of terminals. Similar techniques to that described above may then be used. For example, a terminal could connect itself to a free time slot within the digital bit stream in order to set up a communication path to the called terminal.

6 INTRODUCTION TO SWITCHING SYSTEMS

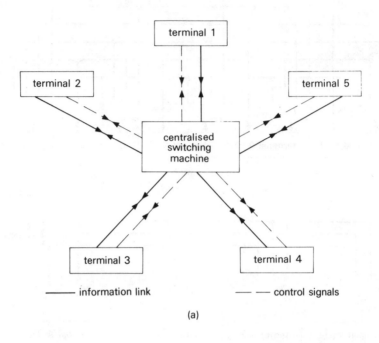

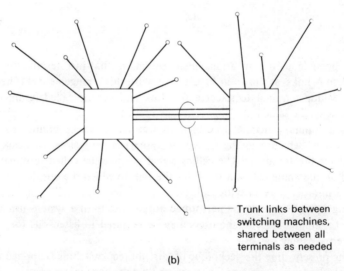

Figure 1.2 The use of centralised switching machines. (a) Single machine which reduces average length of transmission link as compared to Figure 1.1. (b) Use of additional switching machine to reduce transmission costs further (if the terminals have a low utilisation).

Centralised switching centres. In most practical switched networks it is usually more economic to provide a link between each terminal and a central location and to perform all switching operations there, as illustrated in Figure 1.2a. This arrangement significantly reduces the total transmission costs in the network. However, the switching centre must be operated by remote control from the terminal and this tends to increase the total switching costs.

The total transmission costs may be reduced even further if a number of local switching centres are used instead of one national centre because this reduces the average length of the connection between a terminal and its nearest switching centre (Figure 1.2b). The local centres must be interconnected by transmission links which are usually called *trunks*. These trunks are shared by all the terminals connected to each centre, and as will be shown later, the number of trunks connected to a local centre can be very much smaller than the number of terminals.

The use of multiplexing techniques for long distance transmission makes cost per unit distance for a trunk less than the cost per unit distance for the link between a terminal and its local switching centre. Therefore increasing the number of switching centres lowers the total transmission costs. However, as the number of centres is increased the total switching costs tend to increase for two reasons. First, the local centres become more complex because they must be able to decide on a suitable routing to another centre and because the centres involved in a call must be able to exchange information. Secondly, economy of scale is lost with an increased number of local centres because two half-size centres, plus their buildings and power supplies, cost more than one full-size centre. In general there is an optimum number of local centres for minimum total cost of transmission and switching. This optimum number of local centres depends upon the relative costs of switching equipment and transmission equipment and the geographical distribution of terminals.

If certain assumptions are made about the geographical distribution of terminals, their traffic characteristics, and the costs of switching and transmission, it is possible to make mathematical analyses of the minimum cost for a total system [3] (see Appendix A). However, these analyses have limited value because the practical details of a given situation invalidate the generalised assumptions. For instance, in the telephone field, detailed costings of possible network plans are needed in order to decide how many switching centres should be installed to cover a particular area, or, more usually, to decide whether it is better to install a new switching centre in the suburbs of a growing town, or to extend the main centre [4]. These costings are simplified by using detailed computer models of the area under consideration to facilitate rapid estimation of the costs of alternative arrangements.

Hierarchical systems. As the number of separate switching centres increases the number of different trunk routes between them increases. Above about ten centres the number of trunk routes becomes very large and routes tend to contain too

8 INTRODUCTION TO SWITCHING SYSTEMS

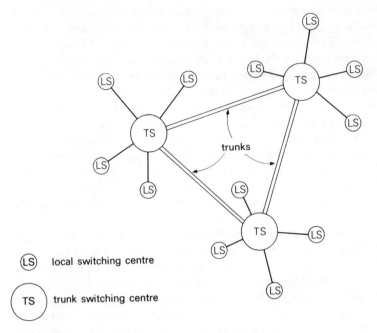

Figure 1.3 Mesh of trunk routes.

few circuits to make the network economic. The same argument that led to centralised local switching now applies to the trunk routes, so trunk switching centres are introduced to switch between trunks. Figure 1.3 shows how the trunk centres reduce the total length of trunk circuits. However, this arrangement introduces the additional cost of trunk switching centres and also necessitates three switching points (rather than one or two) on some connections.

The process of centralising switching centres can occur at several levels leading to what is called a *hierarchical network*. This is best explained by reference to the public telephone network. In a country there is normally a number of local switching centres, each serving anything from 20 to 10 000 terminals (or even up to 100 000 in special cases). These local centres are gathered into groups and each group is served by a trunk centre which can connect calls between local centres within the group. For calls between terminals of one group of local centres, and those of another, the trunk centres themselves have to be interconnected by 'super' trunk centres each covering an area consisting of many local groups. Usually there is a number of these wider areas and these too will be interconnected by higher level trunk centres. This principle is extended to the level where the number of trunk centres at that level is small enough for it to be practical to interconnect them fully with a mesh of trunk routes similar to those shown in Figure 1.3. The result is as shown in Figure 1.4a.

CENTRALISED SWITCHING SYSTEMS 9

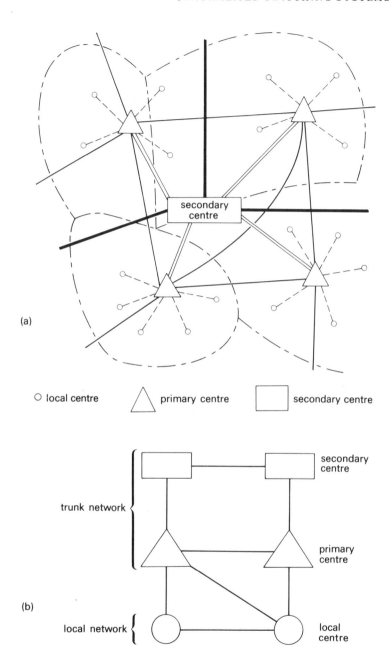

Figure 1.4 (a) Hierarchial network of switching centres. (b) Conventional representation.

Table 1.1 Names of different level switching centres in telephone networks (The numbers in brackets indicate approximate numbers in each country (1976).)

Function	International name	U.K. name	North American name
Local switching centre directly connected to terminals	Local Exchange	Local exchange (6200)	End office or Class 5 office (18 000)
First level trunk switching centre	Primary centre	Group switching centre (370)	Toll centre or Class 4 office (1500)
Second level trunk switching centre	Secondary centre	District switching centre (27)	Primary centre or Class 3 office (250)
Third level trunk switching centre	Tertiary centre	Main switching centre (9)	Sectional centre of Class 2 office (65)
Fourth level trunk switching centre	Quaternary centre	(not required in U.K.)	Regional centre or Class 1 office (18 in U.S.A., 2 in Canada)
Centre which provides connections only between local centres (i.e. no access to trunk network)	Tandem exchange	Tandem exchange	Tandem office
Switching centre which provides access to international network.		Centre du transit 3 (or CT3)	
International transit centre		Centre du transit 2 (or CT2)	
Top level international transit switching centre (fully interconnected)		Centre du transit 1 (or CT 1) (there are 7 of these in the international network)	

In the telephone system, different countries have adopted different names for these different levels of centre. So to simplify discussion a set of names has been internationally agreed [5]. Table 1.1 shows these names together with equivalent U.K. and North American terms.

Figure 1.4b shows a conventional way of illustrating a hierarchical network plan. It can be seen from this that a hierarchical network, as described above, guarantees that a connection between any two terminals will be possible. Also it sets a limit to the number of links required in the worst case. In practice the situation is more complex. Large numbers of direct routes are provided between switches if the traffic level is justified. So, for example, there may be direct routes between a tertiary centre in one area and a secondary centre in an area served by a different tertiary centre.

Functions of telephone switching systems. Some of the functions of telephone switching will now be defined. The local switching centre must react to a calling signal from a terminal and must be able to receive information to identify the required destination terminal. It must be able to decide from the input information whether the required terminal is connected to the same local centre or whether a trunk connection is necessary via one or more intermediate trunk centres. If an intermediate trunk centre is needed the local centre must find a free trunk on the required trunk routes and connect the terminal to it. Further information must then be forwarded to the intermediate trunk centre or centres to progress the call to its destination.

Once a path has been set up from the originating centre to the terminating centre, the called terminal must be rung; and once the called party has answered, a speech path must be established between the two terminals for as long as the call lasts.

Since public telephone systems must make money, at some stage it is necessary to extract charging information for billing purposes.

A further requirement which is not obvious from what has been said so far is that a telephone system must be very reliable. In the language of reliability mathematics, a telephone system must have a *high availability*. Most switching systems are required to give uninterrupted service for many years. Telephone systems, for example, have design lives of from 20 to 40 years. Present technology is such that no system can be guaranteed to be completely free of faults for this length of time, but it is nevertheless possible to design a system to provide an adequate service even in the presence of faults or malfunctions.

System reliability can be expressed mathematically in terms of *availability* defined as:

$$A = \frac{\text{up-time}}{\text{up-time} + \text{down-time}}$$

where the up-time is the total time that the system is operating satisfactorily and the down-time the total time that it is not.

An alternative and equivalent definition of availability is in terms of the mean time between failures (m.t.b.f.) and the mean time to repair (m.t.t.r.):

$$A = \frac{\text{m.t.b.f.}}{\text{m.t.b.f.} + \text{m.t.t.r.}}$$

12   INTRODUCTION TO SWITCHING SYSTEMS

Table 1.2   Some typical availability objectives for public telephone systems

---

For faults causing complete loss of service for more than 3 minutes and

        affecting only a single terminal   – m.t.b.f. $\geqslant$ 10 years
        affecting 10% of terminals   – m.t.b.f. $\geqslant$ 20 years
        affecting complete switching centre – m.t.b.f. $\geqslant$ 50 years

Total down-time (due to switching centre failures) $\leqslant$ 2 hours in 40 years i.e. overall availability $\geqslant$ 99·9994%

Availability objectives are difficult to define. A fault which upsets the service of only one or a small number of users is less troublesome than a fault which makes a complete switching centre inoperative. The period of time that service is denied is also important. For telephone users, a break in service of only a few seconds (if it occurs only rarely) is not too troublesome, but breaks of fifteen minutes or more are very troublesome. Hence the availability figures must be split into a number of different categories. Some examples are shown in Table 1.2.

It should be noted that the origin of these objectives is largely historical; early systems (step-by-step or Strowger switching systems to be described later) actually achieved availabilities of this order.

1.4   Basic switching centre model

Most of the fundamental principles of switching system design can be understood by considering a single centralised switching centre like that of Figure 1.2a. In addition to a channel for transfer of information, there is also a two-way path between each terminal and the centre for interchange of the system control signals. In fact, in most of the existing systems the same physical channel is used for both purposes. One of the practical problems of system design is the separation of the control signals from the information at the switching centre.

The general realisation of this centralised machine (for circuit switching) is shown in Figure 1.5 where each terminal has its own switch and control signal path to its own control unit. Each switch has outlets to each of the other terminals. Also each control system has access to each of the other control units; this is necessary so that the control unit associated with a calling terminal may test whether the control unit associated with a called terminal is free and, if it is free, busy it.

Signal exchange diagrams.   The first step in any system design is to consider the range of control signals that has to be interchanged between a terminal and the system. This information is conveyed in the form of *signals* and many different ways are used to code these signals. In telephone systems a commonly used form of signal is the changing of the value of an analogue quantity, such as

BASIC SWITCHING CENTRE MODEL   13

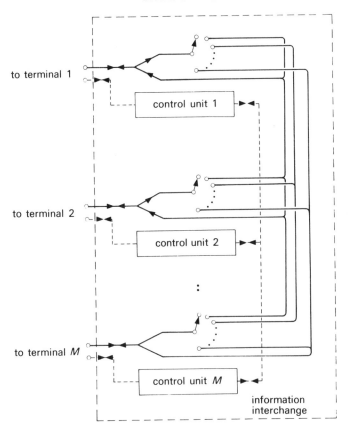

Figure 1.5   Simple model of a centralised switching machine.

the d.c. resistance across a pair of wires, the amplitude of a sinusoidal voltage etc. In message and data systems, special bit patterns are used as signals. Specific techniques are discussed later.

Whatever the techniques used, the basic signal types remain the same. Figure 1.6 (called a signal exchange diagram) shows a basic set of signals for a terminal that can be used in either a calling or a called mode. In the calling mode the first action that the user of the terminal must take is to transmit a *seize* signal to the system to indicate that the terminal wishes to make a connection or pass a message. The system generally responds to the *seize* by an *accept* signal (for reasons which will become apparent later). The terminal then transmits *routing* signals and the system responds with a variety of *status* signals, for example:

- *line busy*   — *number invalid*
- *line free*   — *line answered*

Once the need for the connection is over, the terminal sends a *clear forward* signal. 'Forward' and 'back' here refer to the direction of traffic which is deemed

14 INTRODUCTION TO SWITCHING SYSTEMS

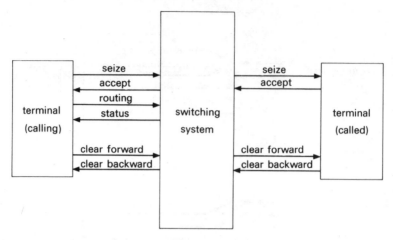

Figure 1.6 Basic signal exchange diagram.

to flow from the originating terminal to the terminating terminal.

In the called mode the terminal is sent a *seize* signal from the system and responds with an *accept* signal. At the end of the period of communication the terminal may be sent a *clear forward* signal to indicate the end of the connection.

In some systems a *clear back* signal is sent from the called terminal to the system and possibly also from the system to the calling terminal if the user of the called terminal is the first to indicate that the communication is finished. For example, the signal sent from a telephone terminal to the local centre when the handset is replaced at the end of a call is regarded as a *clear forward* signal if the user initiated the call, but a *clear back* signal if the user was the recipient of the call, even though the signals may take the same electrical form in practice.

State transition diagrams. The signal diagram gives the 'alphabet' of the valid signals between two devices. However, it does not indicate what sequences (or 'sentences') are possible or what they mean. The valid sequences and their meaning can be expressed conveniently in what is called a *state transition diagram* (s.t.d.). This is such a useful specification and design tool that a set of international standards have been produced [6] for use in telecommunications systems.

The concepts underlying an s.t.d. can be explained by reference to the control units in Figure 1.5. These control units may be thought of as existing in a number of stable states such as:

— *idle* — *waiting for answer*
— *waiting for routing information*

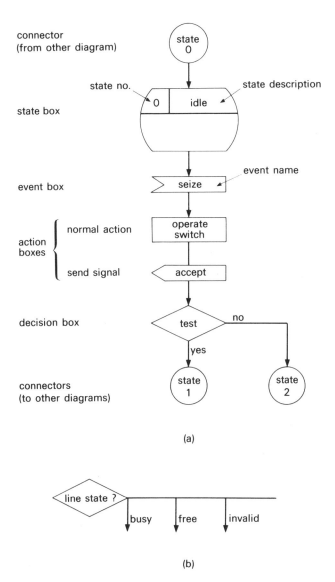

Figure 1.7 Example of state transition diagram symbols. (a) Basic symbols. (b) Multiple decision box.

In general a control unit moves from one state to another only because of the arrival of a signal from a terminal or another control unit. The arrival of a signal is described as an *event*. When the unit does move from one state to another it may perform some action such as operating a switch or sending a signal to

16   INTRODUCTION TO SWITCHING SYSTEMS

another control unit or terminal. The actions performed between one stable state and another are collectively called a *task*.

In many cases the combination of current state and new event defines the task and new state, but in some cases there is a choice of next states. Such a choice depends upon information external to an individual control unit. The most obvious example of this is the choice of next state after the *routing* signal has been received. A different state is necessary depending upon whether the called terminal is busy or free.

Thus there are four components of a state transition diagram, the symbols for them are shown in Figure 1.7:

(a) *State boxes*. These are labelled with a number and descriptive title. In some cases it is useful to place a pictorial representation of the state in box.
(b) *Event boxes*. The possible (or permitted) events are each drawn as arrow-indented boxes, the paths to which come from the state box in question.
(c) *Action boxes*. Actions are shown by rectangular boxes except for the action of sending a signal to another control unit or terminal which is given the special symbol of an arrowed box.
(d) *Decision boxes*. For binary decisions the symbol is the diamond-shaped box as normally used in computer flowcharting. For multiple decisions the extended version shown is used.

S.t.d. for the calling states of a simple control system.   Figure 1.8 shows a simplified s.t.d. for the states of a control unit involved on an originating call. In the *idle* state the only valid event is *seize*. This is acknowledged by an *accept* signal and the control system moves into state 1, *waiting for routing*. The event of reception of *routing* leads to a test of the state of the required terminal. (The mechanisms for performing this test are discussed later.) If the called terminal is free, the calling terminal is informed, a path is set-up, and the unit moves into state 2, *waiting for answer* The *accept* signal from the called terminal causes an *answer* signal to be sent to the calling terminal, and the control unit moves to state 3, the *talking* state.

In this simple example it is assumed that the connection is under the control of only the calling terminal. This arrangement is referred to as calling party release. Any *clear back* signals from the called terminal are therefore irrelevant. There are three other possible call clear down techniques:

> Called party release;
> First party release;
> Last party release.

In states 1 and 2, it is possible for the user of the calling terminal to abandon the call. If this happens a *clear forward* signal is sent to the system. The s.t.d. in Figure 1.8 shows the action to be taken in these cases.

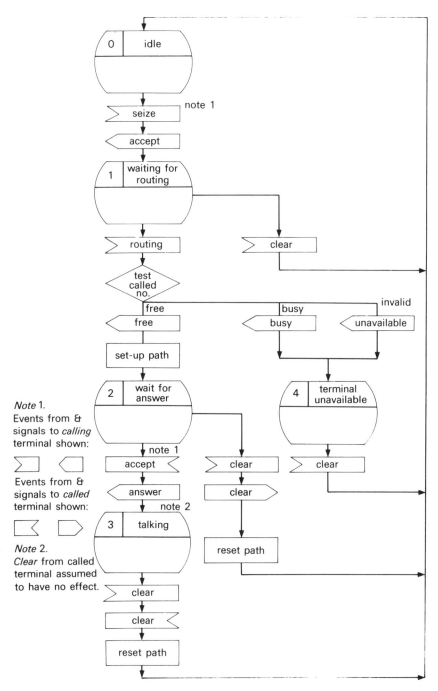

Figure 1.8 Simplified s.t.d. for originating calls only.

18 INTRODUCTION TO SWITCHING SYSTEMS

The complete s.t.d. for originating and terminating calls are described in Chapter 8.

1.5 Resource sharing

A system like the Basic Switching System described above has two inherent advantages:

(a) It can be made with very high system availability, because if it is properly designed, any fault in a control unit or its associated switch should affect at most, only two terminals, that is its own and any one which happens to be connected to it.
(b) In the absence of faults, the only possible reason for an on-demand connection to another terminal not being achieved is that the required terminal is already busy.

With present day technology, however, a system like this is generally uneconomic. The art of switching systems' designers over the last century has been to find techniques to reduce the cost of a switching system, while maintaining high levels of system availability. Practical system designs take advantage of the

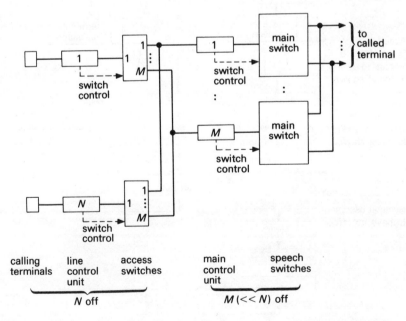

Figure 1.9 Example of functional subdivision. *Note:* lines between blocks carry both control signals and information.

fact that not all parts of a complete system are required all the time, either because terminals are not in use or because they can operate much faster than needed. These factors permit the possibility of *resource sharing* as a means of cost reduction. In switching systems there are three techniques of resource sharing.

(a) Functional subdivision. It is not necessary for all the functions of a control unit to be available all the time. For instance, while it is in the *idle* state a control system for one terminal need only have the function of detecting a *seize* from its terminal or from another terminal that is its calling terminal. Also, when a terminal is in the *conversation* state its control system does not need the functions involved in receiving and decoding the *routing* signals. The technique of functional subdivision involves partitioning a control unit into a number of units, each of which provides only some of the functions of the complete control system. Each terminal needs a permanently associated unit to detect *seize* signals, but other units may be pooled and switching arrangements introduced so that they are associated with a terminal only while they are needed. This technique is shown in Figure 1.9 where the control unit has been partitioned into a line control unit and a main control unit. There is one line control unit per terminal and these have access to a lesser number of main control units.

This technique can reduce the total amount of control equipment, but it introduces the need for extra switching and its control. It also introduces the possibility of a terminal having to wait for service if a common control unit is not immediately available.

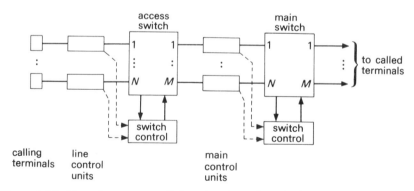

Figure 1.10 Use of common switch networks and controls.

(b) Common switch network and control. The second technique of resource sharing is replacing a number of one-by-N switches by a single multiple-input, multiple-output network. Figure 1.10 shows this applied to the system of Figure

1.9, where the M one-by-N main switches are replaced by a single M-by-N network, and the N one-by-M access switches are replaced by an N-by-M network. It will be shown in Chapter 7 that this considerably reduces the total number of switch elements required in a system. There are two problems with this technique:

(i) In most designs of switch network there is a particular probability that no path will be found between an inlet and outlet, even though the terminals themselves are free.
(ii) It is usually necessary to have a common control for the switch network, and if this goes wrong it will affect the complete system.

(c) Time-Shared Decision Making. Each control system exists in a number of states and changes its state only when some event occurs. When an event occurs its control system must decide on the necessary actions and the new state. The third method of resource sharing is possible because the time between occurrence of events at a control system is much greater than the time required for making decisions. For instance, in a telephone system the shortest time between events for a single terminal is typically 30 ms or more, whereas electronic circuits can make a decision in 1 μs or less. The decision-making component may therefore be time-shared between a large number of control units of a given type. Each unit needs a means of storing its current state, buffering the received event, and storing the required actions.

The method by which these techniques of resource sharing are achieved are discussed in Chapter 3.

Objectives of System Design. All practical switching systems achieve their economic objectives by different combinations of the above three techniques of resource sharing. They all introduce the possibility that a resource, such as a control unit, or a path through the network, may not be available when needed by a terminal. When this occurs, the terminal must either abandon its call or wait until the resource that it needs becomes available. A switching system must provide sufficient resources so that the probability of a resource not being available when it is needed, is kept below some design maximum. A significant part of switching system design is therefore concerned with statistics.

It will be seen later that the implementation of these three techniques introduces system availability problems. The Basic Switching System can be the basis of a highly reliable system structure, but when resource sharing is introduced the possibility arises that a single fault may affect more than one terminal and in some cases, the complete system. One aspect of the art of system design is to develop techniques of resource sharing that minimise the effects of the faults that will inevitably occur. Because a switching system must be economic, the problem of system design is ultimately an economic one.

The problem of switching system design may be summarised as follows:

To develop the optimum level and arrangement of resource sharing which satisfies the functional requirements of the systems and keeps below a design objective the probability of a terminal not being able to set-up a call due to
(a) lack of resource when needed, or
(b) faults within a sub-system.

1.6 Introduction to traffic and queueing theory

The introduction of resource sharing into a telecommunications switching system means that sometimes when a call request is made, the resources required will not be immediately available. In these circumstances a call is said to be *blocked*. For some applications it is assumed that, if a call is blocked, the call is abandoned or 'lost'. In other applications the call request is queued in some way until the required resources become available, that is the call 'waits' or 'queues'.

In practice most systems have a mixture of call loss and call queueing. In a telephone system the caller must wait for dial tone but thereafter, if a required resource such as a trunk line is not immediately available, the call is lost and the user receives a busy tone. Usually, even if the resource subsequently becomes available, the busy tone will not be removed.

Traffic theory and queueing theory are used to estimate the probability of occurrence of call blocking and, if queueing is involved, to estimate the statistical distribution of the waiting times of blocked calls. In fact theory is usually used for design rather than analysis, that is to estimate the quantity of resources required in order for a system to meet particular probabilities of loss and queueing.

A theoretical analysis of the performance of a switching system needs two pieces of information as a starting point:

(a) a statistical description of the traffic demands from terminals, and
(b) a set of performance objectives.

Traffic statistics. The traffic demands from terminals may be characterised by the following:

(a) Calling rate. This is the average number of requests for connection that are made per unit time. If the instant in time that a call request arises is a random variable, the calling rate may be stated as the probability that a call request will occur in a certain short interval of time.

(b) Holding time. This may be characterised most simply by the mean time that calls last. In some circumstances (particularly where there is call queueing) the statistical distribution of the holding times is also important. The most commonly used distribution is the negative exponential distribution in which the probability of a call lasting at least t seconds is given by:

$$P(t) = \exp(-t/h)$$

22 INTRODUCTION TO SWITCHING SYSTEMS

where h is the average holding time in seconds. An illustration of this function is given in Figure 1.11 for $h = 100$ seconds. It shows that with this mean holding time there is:

> 50% probability that a call lasts longer than 70 seconds;
> 36% probability that a call lasts longer than 100 seconds;
> 13% probability that a call lasts longer than 200 seconds.

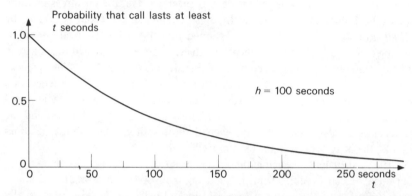

Figure 1.11   The negative exponential function, $\exp(-t/h)$, for $h=100$s.

This distribution is found to fit a large number of practical cases such as local calls in a telephone network. In many cases the holding time for long distance calls also follows this distribution but with a different mean. This distribution has the advantage that it leads to convenient analytic equations. The main reason for this is that the probability per unit time of a call ending is constant, and therefore independent of the length of time that the call has already been in progress.

In the analysis of control systems it is sometimes more appropriate to assume a constant holding time.

(c) Distribution of destinations.   This is described as the probability of a call request being for particular destinations. This is obviously of importance in a hierarchical, many centre system, since it determines the number of trunks needed between individual centres.

(d) User behaviour when no resource is immediately available.   The statistical properties of the switching system are a function of the behaviour of users who encounter call blocking. For instance, the system will behave differently if the user

- abandons the request,
- makes repeated attempts to set up the call,
- waits until the resources become available.

Macroscopic traffic statistics: the erlang. As far as the switching elements of a switching machine are concerned, it makes no difference whether a terminal makes ten two-minute calls or one twenty-minute call. It will of course make a difference to the control because to set up ten calls means more work. For this reason studies of switching networks and their interconnecting trunks are normally made on the basis of total occupancy rather than the two parameters of calling rate and holding time. The internationally agreed parameters used are as follows [7]:

The *amount of traffic carried* (by a group of circuits or a group of switches) during any period is the sum of the holding times expressed in hours.

The *traffic flow* (on a group of circuits or a group of switches) equals the amount of traffic carried divided by the duration of the observation, provided that the period of observation and the holding times are expressed in the same units. Traffic flow calculated in this way is expressed in *erlangs*.

Thus the amount of traffic has the units of time and traffic flow is dimensionless.

The erlang is named after A. K. Erlang, a mathematician who developed much of the early theory of telephone traffic. Applying the definition to an individual terminal, if the average number of calls arising in time T is n and the average holding time of calls is h, the amount of traffic carried in time $T = nh$ units of time and the traffic flow $A = nh/T$ erlangs.

In the case of a single terminal the traffic in erlangs is equal to the average occupancy of the terminal, where occupancy is defined as the proportion of the time that a terminal is busy. It can be easily shown that the traffic in erlangs from a group of terminals is numerically equal to both of the following:

(a) The average number of concurrent calls.
(b) The average number of calls which originate during the average holding time.

An alternative unit for traffic measurement in common use in North America is the *hundred call seconds* (ccs). This is used as a measure of the amount of traffic expressed in units of 100 seconds. The number of ccs per hour is also used as a measure of traffic flow. Since 1 erlang may be regarded as an average of one call for one hour, then:

$$1 \text{ erlang} = 36 \text{ ccs h}^{-1}$$

Busy hour traffic. The actual values of all parameters of traffic, and in particular the calling rate, vary with time. In a telephone system the number of calls made per hour varies throughout the day. For instance, there is a peak in the calling rate in the morning for a telephone system with mainly business users. For a system in a largely residential area the peak may occur in the early evenings. Although the evening peak may be partly due to the operation of a 'cheap rate' tariff, the morning peak is largely due to the habits of business users and is affected only temporarily by the introduction of higher 'peak rate' tariffs. So, cheap rate tariffs may be said to stimulate use of the telephone whereas peak rate tariffs are simply a way of maximising revenue in an acceptable way. For long distance traffic the peaks for different parts of the network may occur at different times due to time-zone differences. Other significant traffic patterns may arise from regular commercial events, for example traffic peaks between fishing ports and London may follow the tides because the fish market follows the landing of the catch.

In addition to changes within the day, the calling rate may vary with the season of the year, for instance it may be higher during the summer season in a holiday resort. There are very large peaks at holiday times such as Christmas, or, in the U.S.A., Mothers' Day. Most switching systems grow with time, in terms of the number of terminals, so the total traffic increases. It is interesting to note that, as a system such as the telephone system grows, there is a tendency for calling rates (per terminal) to decrease during certain stages of this growth, presumably because an increased number of terminals means fewer people sharing each terminal. At other stages, this effect is outweighed by increased use per person, due to the increasing number of easily contractable correspondents.

In order to design a suitable switching system some traffic figures are needed which represent the average demands made by users on the system over a planning period of the order of six months. In the telephone field, the so-called *busy hour* traffic figures are used for planning purposes. An hour is chosen because it is long compared with the average holding time of around 3 minutes. If the parameters are measured for the busiest hour on a number of the busiest days of the year, the mean of these measurements gives a useful measure of the traffic.

Once the statistical properties of the traffic from or to a set of terminals are known, it is necessary to state an objective for the performance of a switching system. This is done by specifying a *grade of service* (g.o.s.). For a system designed on a loss basis, a suitable grade of service is the percentage of calls which are lost because no equipment is available at the instant of the call request. In a waiting system a grade of service objective could be either the percentage of calls which are delayed or the percentage which are delayed more than a certain length of time.

This grade of service is applied to a terminal-to-terminal connection, but in a system containing many switching centres it is usually more convenient to break the objectives down into component parts such as the grades of service for:

— an internal call,
— an outgoing call to the trunk network,

— the trunk network itself,
— a terminating call.

The reasons behind the choice of objectives are not clear cut. They must be a balance between economics and user satisfaction. The economics can be computed, but the user's satisfaction cannot. One objective approach for a commercial system is to find the grade of service for which the cost of adding the extra equipment is equal to the revenue that is being lost by calls being lost [8]. Although easy to state, this computation is difficult to perform and often gives results which, although 'economic', are unacceptable to the user. For instance, if the grade of service for a long distance route is greater than 10% the user will experience considerable annoyance.

Typical objectives for overall grades of service for a commercial telephone system are 3–5% for local calls and (because the cost of equipment provision is higher) up to 6 or 7% for long distance calls. These figures are comparable with the probability that the called user is already busy.

Typical objectives for component parts of a connection are:

Internal Calls 3%	Trunk Calls 1–3%
Outgoing Calls 2%	Incoming Calls 2%

For the above, the overall grade of service is in fact approximately the sum of the component grades of service.

However, this is not a complete specification because it relates only the grades of service for a mean busy hour. In order to ensure that the grade of service does not deteriorate disastrously if the actual busy hour traffic exceeds the mean (or if some of the equipment is out of service), additional grades of service are specified relating to traffic loads of 10% or 20% above the mean, or with certain percentages of the system not operational.

Traffic design objectives. Traffic design objectives for a switching system are generally expressed in terms of a set of traffic flows to be carried by the system with a performance no worse than a set of grades of service for the different types of traffic. In addition, the performance of the system must not deteriorate beyond a certain set of grades of service under specified overload or fault conditions, such as 10% overload or only 90% of the equipment being available because of faults.

The grades of service that are relevant for a message switching system relate to delay. All communications are delayed to a certain extent so a system might have an objective of the form: 99% probability of delivering a message within 24 hours of transmission. This may be too slow for certain types of message, so a range of message priorities is normally used to allow the more urgent messages to 'jump' the queue and be delivered ahead of previously transmitted, lower priority messages. The existence of these levels of priorities complicates the analysis.

A data system may be either circuit switched or packet switched. In a circuit switched system conventional loss probabilities may be used. The critical factor in a packet switching system is usually response time, for example the time taken for a packet to travel from one terminal to a main computer, for that packet to be processed, and for response to be returned to the originating terminal. If there is a human user, acceptable performance standards are typically a mean response time of 1·5 seconds and 90% of responses obtained within 3 seconds [9].

References

1. Hills, M. T. and Evans, B. G. (1973) *Telecommunications Systems*. Allen & Unwin.
2. Beck, I. H. (1972) Mobile radio systems, *Post Office Elect. Eng. J.*, **64**, p. 238.
3. Rapp. Y. (1950) The economic optimum in urban telephone network problems, *Ericsson Technics*, 49.
4. For example, see Back, R. E. G. (1975) Network planning, in *Telecommunications Networks*. J. E. Flood, ed., Peter Peregrinus.
5. *CCITT (1964) National Telephone Networks for Automatic Service*. ITU.
6. *CCITT Orange Book* (1977) Vol. VI, Recommendation Z101.
7. *CCITT Orange Book* (1977) Vol. II, Recommendation E160.
8. Jenson, A. (1950) *Moes Principle. An Economic Investigation Intended as an Aid in Dimensioning and Managing Telephone Plant*. Copenhagen Telephone Company.
9. Martin J. (1972) *Systems Analysis for Data Transmission* (Chapter 7). Prentice Hall.

2
Signalling and switching techniques

2.1 Telephone systems

The terminal and local signalling. The terminals of a telephone network are the familiar telephone instruments. There are many different styles of instrument throughout the world, although electrically these are all basically similar. A simplified circuit of the transmission elements of a telephone terminal is given in Figure 2.1 [1]. The microphone is almost always a carbon microphone and therefore needs a direct current to power it. This is normally provided by a d.c. power supply (conventionally referred to as the 'central battery' in the local switching centre). The d.c. is fed through what is called a *feeding bridge*. The feeding bridge is in effect a filter which separates the d.c. (a 'zero hertz' signal) fed to the terminal but passes audio frequencies in both directions. A transformer is used in the terminal to match the low impedance of the microphone to the line and to reduce what is called 'sidetone'; this is the sound of the user's own voice caused by speech energy from the transmitter reaching the receiver of the same instrument [1].

The terminal is usually connected to the local switching centre by a pair of wires which is electrically balanced about earth and carries the speech signals in both directions. The balancing prevents interference between telephones whose pairs are adjacent to one another in cables or on telegraph poles. It also minimises the effect of induced currents produced by nearby power cables and other forms of interference. This pair of wires is called the *subscriber's loop* in North America, and the *local line* in the U.K.

The control signals required for the remote control of the switching system are commonly transmitted on the same pair of wires as the information. This is an obvious saving in transmission cost, but it implies that suitable equipment must be provided at the switching machine to extract the signalling information received from the terminal and to insert signalling information sent to the terminal.

28 SIGNALLING AND SWITCHING TECHNIQUES

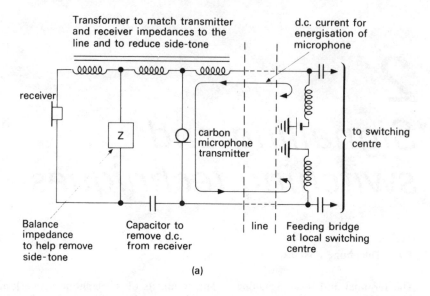

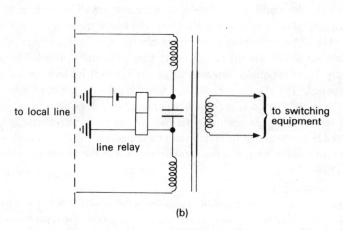

Figure 2.1 Simplified circuit of telephone instrument from transmission viewpoint. (a) Basic principle. (b) Alternative form of feeding bridge.

The basic signals of *seize, answer* and *clear* are provided by changes of the d.c. resistance connected across the pair of wires at the terminal. When the terminal is not in use, the handset rests on a cradle which operates a switch. This is called the *on-hook* condition. Removing the handset from its cradle operates the cradle switch and connects the transmission circuit across the line. The change from open circuit to closed circuit therefore represents the *seize* signal, or, if the telephone is being called, the change represents the *accept* signal. In both cases replacement of the handset at the end of a call is the *clear* signal (*clear forward* if it is the calling terminal, *clear back* if it is the called terminal).

These d.c. signals can be detected at the feeding bridge in the local switching centre if the bridge incorporates a relay which operates whenever a line current is flowing. To call a terminal (i.e. send a *select* signal) an a.c. signal is generally used; a bell, in series with a capacitor, is connected across the line at the terminal, so that it rings when a.c. is applied at the switching centre.

In a manual telephone system, the *routing* signal consists of verbal instructions to an operator who operates the necessary switches and gives a verbal report of the progress of the call. In an automatic system the *routing* signal must be machine-recognisable; a number of forms of signalling have been developed. The earliest and still the most common is based on the use of the dial. This is an extremely ingenious and cheap device. The basic idea is to interrupt the d.c. path of the subscriber's loop for a specific number of short periods to indicate the number dialled [2]. This is called *loop-disconnect* (or *rotary*) signalling and in most countries the dial operates at about ten impulses per second with a break of about $66\frac{2}{3}$ ms and a make of about $33\frac{1}{3}$ ms. It is necessary to indicate the end of a pulse train so that a decoding circuit can know where one digit ends and the next begins. This is achieved by the mechanical design of the dial which ensures a minimum make period between any two consecutive digits; this period is called the *inter-digit pause* (i.d.p.) and is typically 200 ms minimum. The *clear* signal is produced when the user replaces his handset, breaking the d.c. path. This break is distinguished from the break produced by dialling by its duration. Any break lasting longer than a few hundred milliseconds is taken as a *clear* signal. The line conditions are shown in Figure 2.2 and the circuit for a complete terminal is shown in Figure 2.3.

The use of a dial to send routing signals to the switching centre is fairly slow and inconvenient when ten or more digits are needed, as with nationwide and international dialling. A more rapid method is designed for use with push-buttons. The principal method uses a pair of tones to signal each digit [3]. These two tones are chosen as follows:

one from 697, 770, 852, 941 Hz and
one from 1209, 1336, 1477, 1633 Hz

as shown in Figure 2.4.

There is a total of sixteen possible pairs of tones. Because only ten are needed for ordinary telephone numbers, the other six may be used for additional control signals. When only ten or twelve buttons are used the 1633 Hz tone is omitted. In practice it is possible to build tone receivers that reliably recognise a pair of tones within 40 ms. Allowing 40 ms inter-digit pause, a maximum signalling rate of about twelve digits/second is feasible. Because there are extra buttons this signalling method has the advantage of allowing the user to signal more complicated requests to the equipment. Furthermore, it is possible for a

30 SIGNALLING AND SWITCHING TECHNIQUES

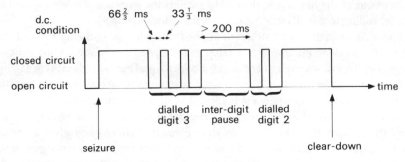

Figure 2.2 Signalling conditions from dial telephone.

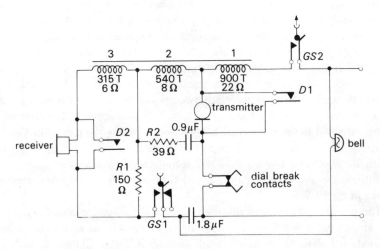

Figure 2.3 Signalling elements of the U.K. Post Office telephone type 706 (courtesy B.P.O.). The $GS1$, $GS2$ switch is operated by lifting the handset from its rest. The dial contacts $D1$ and $D2$ are operated while dialling. $D1$ reduces resistance of telephone circuit by shorting the microphone. $D2$ prevents clicks in the earphone. Note that the 1.8 μF capacitor is used as part of the balance impedance as well as providing d.c. isolation of the bell. It also acts as a spark quench arrangement across the dial break contacts.

user to make a telephone call to a computer, use the tones to send data, and thus obtain stock market quotations, credit rating checks, and so on, with the computer using tape-recorded voice replies (or even voice synthesisation).

Other forms of push-button terminal exist. One form uses an electronic circuit within the instrument to remember the sequence of buttons pushed and to generate loop-disconnect signals for the digits. This permits the use of

TELEPHONE SYSTEMS 31

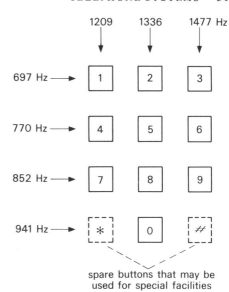

Figure 2.4 Layout and frequency code for push-button telephone.

spare buttons that may be used for special facilities

push-button instruments without the need to change a switching system based on loop-disconnect signalling from terminals. Another form uses different combinations of d.c. conditions between earth and each member of the pair of wires. This latter form has some applications in private automatic branch exchanges because it is cheaper than the tone method, but it is unsuitable for public systems because it is affected by interference from power cables and because it is often difficult to provide a good earth connection at every user's premises.

The signals from the switching system to the user consist mainly of audible tones. The *accept* signal is the start of dial tone. *Terminal free* is the start of ring tone. *Answer* is the cessation of ring tone. Note that it is the start or the cessation of tones of this type which constitutes the signal to the user, that is transfers information to the user.

Trunk signalling. When a call needs to pass through more than one switching centre it is necessary to exchange control signals between the centres. Signals usually include the information supplied by the user but further pieces of signals are useful.

In a switching system the control signals may be categorised into two classes:

line signals which control the setting-up, answering and clear down of the call;
register signals (or routing signals) which are used during the setting-up of the call.

In a telephone system two different signal transmission techniques are often used, one for each class of signal, because line signals may be sent at any time during a call, but register signals are only sent during the set-up phase. Also the amount of information required for register signals is much greater than that needed for line signals. Application of the principle of functional subdivision often leads to the provision of separate sub-systems for line signals and register signals. In this case the line signal sub-systems are permanently connected to a trunk line, whereas the register signalling systems are connected only when required. (Certain line signals indicate when the register sub-system is needed.) Details of some practical trunk signalling systems are contained in Chapter 6.

2.2 Data and message switching systems

In data and message switching systems, the terminal generates and responds to binary data in serial form. For low speed terminals (up to 110 bits/second) it is practicable to use d.c. transmission of the data signal. Teleprinters have traditionally used ± 80 V as the two conditions. At this data rate it is possible to use unbalanced transmission, with one wire of a wire pair used for sending and the other for receiving (with respect to earth). The digital signal from a teleprinter is in the form of a square wave and so has many harmonics of 110 Hz. In order to remove the effects of interference it is necessary to include a low-pass filter at the transmitter.

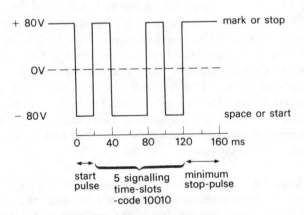

Figure 2.5 Start/stop format for the five-unit code.

At higher speeds balanced transmission is necessary and frequency modulation or phase modulation methods are generally used. As with telephone systems, a major problem is the separation of control from message information.

The output of teleprinters is called *anisochronous* (formerly asynchronous) or, more frequently, *start/stop*. This means that a transmitted character may begin at any time and the sender and receiver have to remain in mutual

Table 2.1 Example of telex signalling

Signal	Direction	Polarity
Line idle	⇄	Start
Call	→	Stop
Proceed to select	←	Stop for 100 ms
Dialling	→	Either 10 pps loop-disconnect or via the keyboard
Clear	→	Start for greater than 300 ms
Connected	←	Stop
Force release	←	250 ms stop followed by 1250 ms start (repeated continuously until line idle detected from other end)

synchronism for only the duration of a single character. The two electrical conditions from a teleprinter are traditionally called 'start' polarity (binary 1) and 'stop' polarity (binary 0). They are also called 'space' and 'mark'.

Figure 2.5 shows the coding technique used for the teleprinter systems in Europe. An idle teleprinter transmits a continuous stop potential and the start of a character is indicated by 20 ms of start potential. The characters themselves are in a 5-bit code which is transmitted in the next five 20 ms time slots as start or stop potentials. The character is ended by a period of at least 30 ms stop potential. The stop pulse has to be of a different length from the other pulses to make it possible to separate individual characters in a continuous stream of characters.

This code uses $7\frac{1}{2}$ units of 20 ms for a single character and the minimum character length is therefore 150 ms. This leads to a maximum character rate of 400 characters/minute. In North America the same codes are used but the unit is 22 ms and the minimum stop duration at the end of a character is 1·4 units. This leads to a slightly slower character rate of 368 characters/minute.

The more modern teletypewriter uses a similar coding but with 8 bits of information per character (including one parity bit) and a full 2-unit stop indicator. The teletype itself typically generates a 110 bits/second signal (600 characters/minute). Other anisochronous devices use the same coding technique but with shorter unit (or bit) durations.

With start/stop coding, continuous start or continuous stop polarity may be used for control signalling purposes in the absence of character signals. Table 2.1 shows how these conditions are used in the U.K. telex network. In the U.K. the routing signals are transmitted by a telephone dial, but elsewhere actual keyboard signals are used. Call progress information is returned as printed characters.

Table 2.2 Example of F31 telegram message format

Item	Start of message signal	Channel sequence numbers			Telegram identification group	End of line
		Channel indicator	Sequence number	Channel indicator	Sequence number	
Number Line	ZCZC	→ GEB	↑ 099	→ AFA	↑ 135	LX→74	<≡

Item	Destination indicator		Priority and Tariff	Origin indicator		Number of chargeable words	End of line
	Country	Office		Country	Office		
Pilot line	→ GB	LO	→ HL	→ UR	WA	→↑ 013	<≡
Preamble line	→ WASHINGTON →↑ 13/12 →↑ 13 →↑ 1205 <≡≡≡						
Paid service indications	→ LT <≡						
Address line	→ MIDBANK <≡↓ LONDON <≡≡≡						
Text	→ FORWARD→SOONEST→PRESENT→ACCOUNT <≡						
	→ BALANCE→JONES→NUMBER→↑78↓A↑765 <≡						
Signature	→ ↑↑↑↑↑ JOHNSON <≡≡≡						
Collation	→ COL→LT→↑78↓A↑765 <≡≡≡≡≡≡≡≡≡						
End-of-message signal	→ NNNN→→→→→→→→						

Key to symbols representing functional combinations:
→ = Space
↑ = Figure Shift
↓ = Letter Shift
< = Carrier Return

In message switching applications the only control signals used are those for line signals between a terminal and its local switching centre. The majority of the control information must be transmitted in the message itself. Table 2.2 shows the format of the messages used for international public telegrams. This is known as the F31 format (the number of the CCITT recommendation) [4]. This format arose from earlier manually controlled systems known as *torn tape systems*. In these a message was printed out at the local switching centre and a paper tape was produced as well. An operator would read the heading in the message to discover the destination. When the message was finished (indicated by the NNNN combination) the operator would tear off the tape and feed it into a transmitter connected to the appropriate outgoing route. If the transmitter was already in use, the new message had to wait. These torn tape centres were replaced by computers, and this implies that the computer must examine each received character to determine whether it is part of the control information or the message.

In the case of teletypewriters with 7 information bits, sending control signals is simpler since specific 7-bit codes can be reserved for control functions. For instance, the American National Standard Institute have standardised a 7-bit code called ASCII (American Standard Code for Information Interchange). In this code, if bits 6 and 7 are both zero, the remaining 5 bits are regarded as control characters. The main disadvantage of this arrangement lies in the transmission of 'general' (non-text) data. In general data, all possible combinations of 7 bits could occur and some might simulate control signals. The system is then said to lack *transparency* because there must be some artificial restriction on the bit patterns to be transmitted. There are various ways around this problem. One common technique is to use a specific control character to disable the switching centre's search for further control characters. It is then possible to send general data, but at least one character must be reserved to indicate return to normal mode. Since this character could be one of the general data items, a special way of sending this character as data has to be used.

This approach is inelegant but practical. A more recent development in what are called *Data Link Control Procedures* is the high level data link block format whose principle is shown in Figure 2.6 [5]. This technique is not tied to any specific character size or transmission method. It transmits information in *frames* which are delimited by the specific bit sequence 01111110, called a flag. Control signals then occupy a specific number of bit positions following a flag (and also preceding it at the end of a frame). In order to protect the system from message or control data inadvertently simulating the flag sequence, the transmitter monitors the bit sequence to be sent and if it finds a sequence of five ones it inserts a zero. At the receiving end, once the flags and control information have been removed, the remaining stream is examined and any zero preceded by five ones is deleted. This guarantees complete bit sequence transparency.

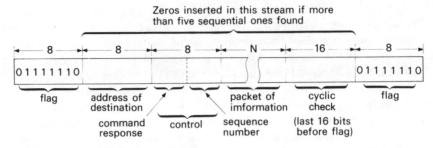

Figure 2.6 Format of high level data link control (hdlc).
*The control field contains indicators to verify correct reception of previous packets of information.

2.3 Switching elements

Types of signal to be switched. A pair of wires which can carry signals in both directions at the same time is referred to as a *two-wire circuit*. In order to provide a means of connecting this type of circuit to other local lines or trunks, each switch in the local switching system must be capable of connecting at least two wires to two other wires and must be capable of transmitting signals in both directions. It would be possible to have only one speech wire to be switched if, within the switching centre, the signal were converted from a balanced to an unbalanced form by means of a transformer as shown in Figure 2.7 [6]. However, the use of unbalanced signals within the switching centre can lead to interference being picked up by other unbalanced transmission paths within the centre, thus increasing the background noise and possibly allowing conversations to be overheard by other users. (This is called crosstalk.) In general, the use of unbalanced switches is only practicable in very small switching centres where the physical length of the unbalanced path is sufficiently small to avoid significant pick-up.

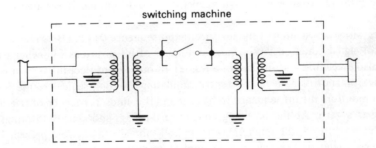

Figure 2.7 Use of balance-to-unbalanced transformers to provide single-wire transmission within a switching machine.

SWITCHING ELEMENTS

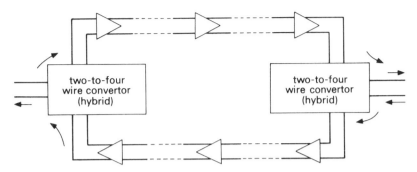

Figure 2.8 Use of two-to-four wire converters (hybrids) in trunk transmission systems.

Within a switching system it is often more economic to transmit the control signals along a separate path from the information. So each switch may have to switch more than two wires.

The trunks used in long distance telephone networks are usually provided by multiplexed transmission systems. It is necessary to split the two directions of speech to produce what is called a *four-wire circuit* (Fig. 2.8). Before the circuit can be switched at the other end of the trunk it often has to be reconverted to a two-wire circuit. As is explained in reference [1], there is in fact a considerable advantage to the overall transmission system if the intermediate trunk switching centres can switch the four-wire circuit directly. The switches in such systems must switch at least four wires, but they need only pass signals in one direction.

In video systems, the high bandwidth involved makes it impracticable to construct two-to-four wire converters and therefore it is not normally practicable to use a cable for bidirectional transmission of video signals. As a result, all video switching must be four-wire if two-way communication is needed.

Data signals in analogue form can be switched by analogue switches. However, in switching systems designed for data signals, the signal is first converted to digital form. This means that the switch elements themselves may be digital and use conventional logic gates.

Types of switching mechanism. There are three classes of switch based on the division of information in space, time and frequency. The difference can be shown by considering the simple example of Figure 2.9a in which three inlets have to be connected to any of three outlets.

Space division. A space-divided switch is one which provides a fixed physical path for the duration of a connection. This could be constructed by three 3-way switches as shown in Figure 2.9b in which, for a particular set of connections, inlet 1 is connected to outlet 2, inlet 2 is connected to outlet 3 and inlet 3 is connected to outlet 1.

38 SIGNALLING AND SWITCHING TECHNIQUES

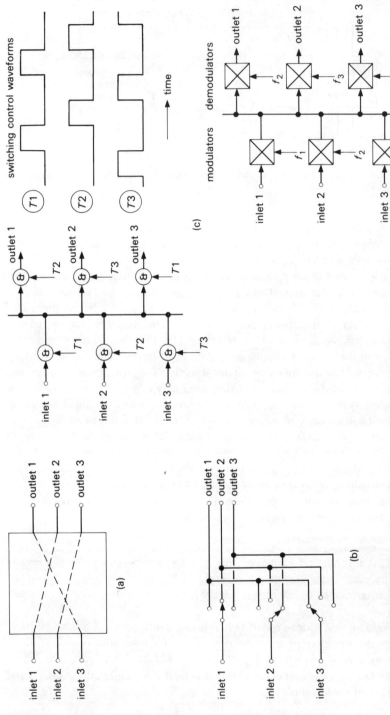

Figure 2.9 Difference between space, time and frequency division switching. (a) Requirement. (b) Space-divided switching. (c) Time-divided switching. (d) Frequency-divided switching.

Time division. An alternative to providing a separate physical path for each connection is to connect all the inlets and outlets to a common transmission medium via high speed switches. These switches connect the required inlet and outlet onto the same path for a short time, followed by another inlet and outlet, as in Figure 2.9c. Each input is in effect 'sampled' by the operation of its switch. This process is repeated periodically. The connection pattern can be changed by altering the time at which the outlets are switched. This technique may only be used where the signal is not affected by the sampling process. For example, it may be used for digital information which has already been sampled or for analogue signals which have been bandlimited before sampling. In the latter case a reconstruction filter (and also usually an amplifier) is needed on the outlets.

Frequency division. A second alternative to space division involves the use of a common transmission path, as in time-division switching, but in this case each signal is modulated onto a different carrier frequency (Figure 2.9d). Switching is achieved if each outlet is provided with a demodulator which can have its carrier frequency changed.

The advantage of space division is its simplicity and the fact that the bandwidth of the signal is not unduly limited by the switch mechanism. It is of course limited by crosstalk considerations and by the characteristics of the switch itself. The disadvantage of space division is that the switches can be relatively slow to operate, can be bulky and can involve a large amount of wiring because each connection needs its own set of wires.

Time division can be fast and compact, but it is usually of practical value only when the signal is already in a digital form. As will be shown later, time-division switching of analogue signals has only limited application.

Until recently, there was no practical application involving frequency division switching, apart from that implicit in radio communication. This is because a space-divided switch is needed to change the demodulator frequencies and so no savings are obtained. However, frequency division switching is now finding application in demand-assigned satellite communication links, and an example is given at the end of this section.

Space-division switches

(a) Conventional relays. These are simple devices which, when used for switching, connect one inlet to one outlet via a set of contacts (Figure 2.10a). In order to build a switching network, relays have to be arranged into matrices as shown in Figure 2.10b. Each relay is known as a *crosspoint*. A space switch must remain operated (or *latched*) for the duration of a call. This latching may be achieved by electrical, magnetic or mechanical means. The principle of electrically latching is shown in Figure 2.11. Magnetic latching can be achieved by the use of combinations of hard and soft magnetic materials in the magnetic circuit. A pulse of current operates the relay and it is released by a pulse of current of

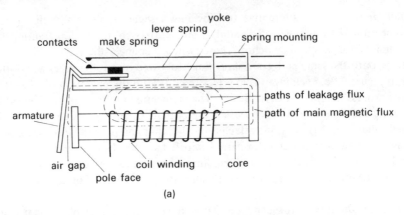

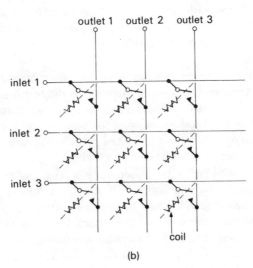

Figure 2.10 Use of relay as a crosspoint. (a) Principle of relay design. (b) Relay contacts arranged in a matrix form.

the opposite polarity. Mechanical latching relays are also sometimes used; these require two solenoids, one to operate and the other to release the contact.

(b) Reed relays [7]. One common electro-mechanical switch used in modern switching equipment is the reed relay. This consists of a pair of contacts made of a soft magnetic material and enclosed in an evacuated or rarefied atmosphere as shown in Figure 2.12a. A number of these contacts may be placed inside a single coil. When the coil is energised the contacts are attracted to each other and a connection is made. The advantage of this type of switch is that it is fast

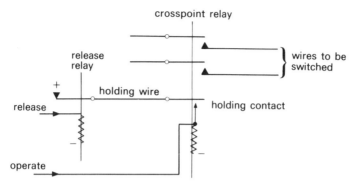

Figure 2.11 Principle of electrically latching relay. The crosspoint relay is operated by connecting an earth onto the operate wire. Once operated, the crosspoint relay stays operated via the holding contact. The crosspoint relay is released by operation of the release relay which removes the earth from the holding wire.

operating (about 2–10 ms), has a potentially long and reliable life and is cheap to construct and handle by automatic means. In switching centres, reed relays are usually arranged in matrices and need to be electrically latched.

An alternative to electrical latching is a magnetic latching reed relay called a Ferreed [8], the basic concept of which is shown in Figure 2.12b. There is one permanent magnet and one magnet whose direction of magnetisation can be reversed by means of a current pulse through its winding. When the magnets are both magnetised in the same direction, the contacts operate. The main disadvantage of this system is the large currents needed to operate the coils and the associated back e.m.f. A more recent development is the remanent reed which relies on the magnetic properties of the reed contacts themselves to provide the magnetic latching [9] (see Ch. 11).

(c) Uniselectors (Fig. 2.13). These are an extension of the principle of the relay in that they can connect one inlet to one of a number of outlets, instead of just one. There are many forms of these devices, the simplest of which uses an electromagnetically operated ratchet to step a number of wiper arms around a bank of contacts. Another version uses a motor with an electromagnetic clutch to stop the wipers at a particular position. A large number may be driven by a heavy-duty common motor with individual selectors operated by means of electromagnetic clutches.

The number of outlets that can be provided on a uniselector depends on the speed with which the wiper can be rotated across all the contacts. For instance, the uniselector shown in Figure 2.13 has a stepping speed of typically 25 steps per second. A motor uniselector can go as fast as 200 steps per second. In most switching machines it is necessary to position the selector within a short time

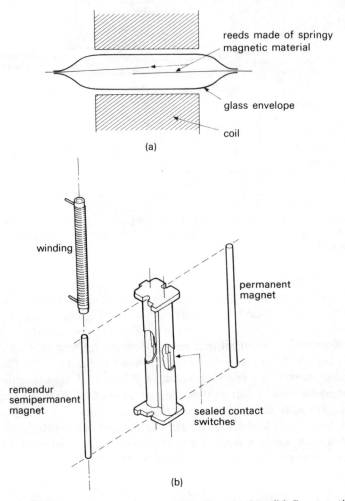

Figure 2.12 Reed relays. (a) Cross-section of reed relay. (b) Construction of a Ferreed — a magnetically latching relay.

(typically 50–100 ms). The number of effective outlets can be increased by means of a technique known as *wiper switching*. A switch with 50 outlets of 3 wires each can be constructed from a 25-way uniselector with 6 wipers, as shown in Figure 2.14. A separate control relay is required to switch the inlet between the first or second set of 3 wipers. The technique may be extended to triple or even quadruple wiper switching. For instance, a 25-way, 12-wiper uniselector may, with the addition of appropriate control relays, be used as:

a 25-outlet 12-wire switch, a 75-outlet 4-wire switch, or
a 50-outlet 6-wire switch, a 100-outlet 3-wire switch.

SWITCHING ELEMENTS   43

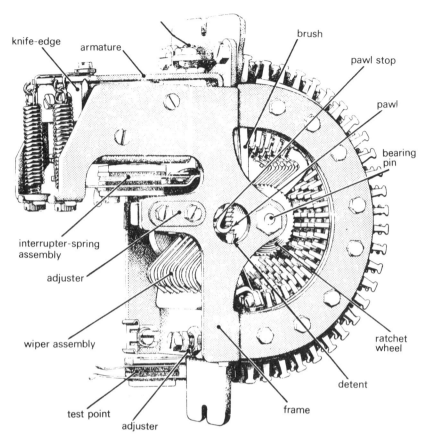

Figure 2.13   Construction of a uniselector.

(d) Two-motion selectors.   A two-motion selector is a selector in which a set of wipers is moved in two different planes by means of separate mechanisms. The best-known example is the Strowger switch, shown in diagrammatical form in Figure 2.15. Typically, the outlets are arranged in banks of ten rows of ten contacts each. A given outlet may be reached by between one and ten vertical steps followed by one to ten horizontal steps. If a number of wires need to be switched, several wipers and several banks are provided. For the design shown, the speed of vertical stepping is limited to ten levels of contacts per second, but the speed of horizontal switching is similar to that of a uniselector.

The major disadvantages of the Strowger type of switch are:

(i) The large number of contacts in each switch makes it uneconomic to make them from precious metal and they therefore need frequent cleaning.

44 SIGNALLING AND SWITCHING TECHNIQUES

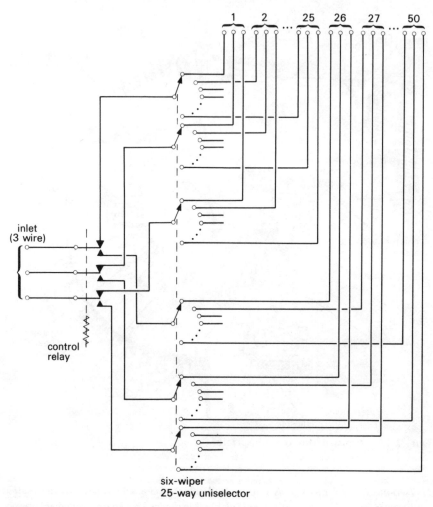

Figure 2.14 Use of wiper switching to increase number of outlets from a switch.

(ii) The large currents needed to operate the switch mechanism.
(iii) The time taken to position the switch contacts.

(e) Crossbar switch. Another type of switch which does not have the disadvantages of the Strowger switch is the crossbar switch and later switches developed from it. This type of switch has a short operation time (less than 100 ms), and is basically a multi-input multi-way switch unlike the two-motion selector which is a *single*-input multi-way switch. A typical switch provides ten inlets which can be connected to twelve outlets of ten contacts each [10].

SWITCHING ELEMENTS   45

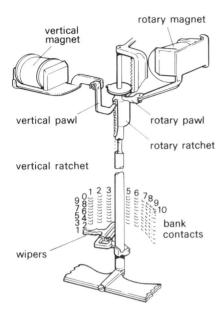

Figure 2.15   Strowger two-motion selector mechanism.

As will be shown in Chapter 7, the use of switches such as a crossbar switch in co-ordinate arrays leads to switching networks which require fewer crosspoints per connection than are required in one- and two-motion selector systems. It is therefore practical to spend more money on the contacts and make them of precious metal (usually silver) to give higher reliability.

The operation of the crossbar switch may be understood by careful study of Figure 2.16. A crossbar switch consists of a number of large relays (typically ten) called bridges, each having a number of spring sets (ten to twelve typically). A bridge differs from an ordinary relay in that it has the ability to select which of the spring sets are to be operated. This selection is performed by a series of select solenoids which are common to all the bridges of the switch. The bridges are normally mounted vertically and a series of select bars are arranged horizontally over the spring sets. These bars can rotate through about 10 degrees, either up or down, by operation of the appropriate select solenoid. Each bar carries one select finger for each of the bridge relays. This finger normally is midway between two of the spring sets. When the bridge solenoid operates, all the select fingers associated with that bridge are pushed between the spring sets and no spring sets are operated. However, if prior to the operation of the bridge solenoid, one of the select bars is rotated, the operation of the bridge solenoid traps the select finger that has been moved between the bridge relay armature. This operates the selected spring set.

The select fingers are made of a springy material, so once the bridge solenoid has operated, the select bar may be released. It then returns to its home position. The chosen select finger remains trapped by the bridge solenoid and that spring

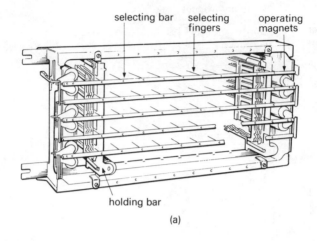

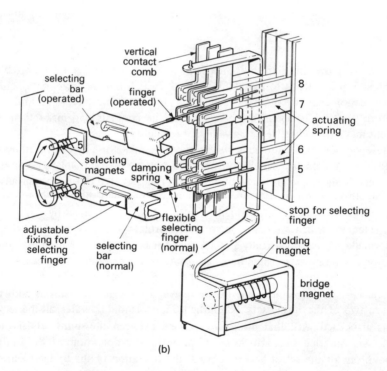

Figure 2.16 (a) Outline of a crossbar switch. (b) A crossbar switch and its operation (detailed mechanism).

set remains operated. The select bar may then be used again if one of the pair of spring sets on another bridge relay needs to be selected. The connection may be released at any time by de-energising the appropriate bridge solenoid. It should be noted that it is possible to make any number of connections (up to the number of select bars) in any one bridge relay.

The number of outlets may be increased at the expense of the number of contacts per crosspoint by wiper switching.

(f) Mechanically latching crossbar switches. The crossbar switch described above is electrically latching and this means that there is a continuous drain on the power supply of the switching machine. In recent years a number of different mechanically latched crossbar switches have been developed, for example the Ericsson code-bar switch [11], the Japanese miniswitch [12] and ITT's miniswitch [13].

(g) Electronic space-division switches. There are many electronic devices that might be used as space-division switches. The basic properties required of such a device are:

(i)   A high 'off' resistance ($>10^8$ Ω) and low 'on' resistance (<1 Ω).
(ii) A low power consumption, especially when 'off'.
(iii) Easy to control.
(iv) Economic.

In addition some form of inherent 'memory' in order to achieve latching is advantageous. If the switch is to be used for analogue message information further required properties are:

(v)   It must be capable of handling the bandwidth and power range of the signal to be switched.
(vi) It must introduce very little noise.
(vii) It must pass signals in both directions if it is to be used in a two-wire switch.

For the switching of telephone signals, no electronic device has yet been found to be suitable for large switching centres. However, recent development of integrated circuit switches based on field effect transistors or p.n.p.n. diodes has made analogue electronic switching a possibility for small systems (e.g. up to one or two thousand terminals) [14]. In these cases though, the low 'on' resistance is not achievable and it is necessary to provide amplification to overcome the transmission loss of the switches.

The cost of most electronic devices is such that only one may be used per crosspoint and so it is necessary to convert from balanced to unbalanced working within the exchange. This means that special arrangements have to be made for signalling. The majority of present day telephone switching systems therefore still use some form of electro-mechanical switch. The most popular switches are various forms of mechanically or magnetically latching crossbar switch and matrices of reed relays.

Time-division switches

(a) Two-wire systems. Figure 2.9d showed the principle of time-division switching. For analogue signals this involves pulse amplitude modulation of the signal during switching. Because the signals are sampled, it is necessary to bandlimit them to a bandwidth equivalent to less than half the sampling frequency in order to prevent distortion. It is also necessary to use a low-pass filter to reconstruct a speech signal from the pulse signals at the receiving end. Provided the switches are bidirectional, it is possible to use the same low-pass filter for both bandlimiting and reconstruction. An example of this type of system as applied to telephone signal switching is shown in Figure 2.17.

At first sight it might appear that this system would introduce a transmission power loss equal to $(\tau/T)^2$ where τ is the pulse width of the sample and T is the period between samples. However, it is theoretically possible to arrange for all the power to be transferred from one filter to the other within the short sampling time by a technique known as *resonant transfer* [15] [16]. The basis of this technique is shown in Figure 2.18. Each line circuit consists essentially of a low-pass filter with a terminating capacitor which acts as a store of energy. In the interval between two sampling pulses this capacitor is charged up to the voltage of the incoming signal. It can be shown that, if there is a particular value of series inductance L associated with each capacitor, the charge on the incoming capacitor can be completely transferred to the outgoing capacitor during the period τ, during which they are connected together. The value of L needed is that which creates a resonant circuit with the store capacitors at a frequency of $1/2\tau$ Hz.

If, when the switch is closed, there is some charge on C_1 the LC circuit starts to oscillate. With the resonant frequency chosen, when the switch is opened τ seconds later, the circuit has executed a half cycle of oscillation. Thus all the energy on C_1 has been transferred to C_2. Similarly any charge on C_2 has been transferred to C_1. The system is thus a lossless two-wire switch. In practice the design has to be modified to allow for stray capacitance on the highway and the ideal, lossless behaviour, is never attained because of the resistive loss of the gates and highways.

A simple version of a time-division switching system may be constructed if there are at least as many time slots as the required maximum number of conversations. It is necessary to have some form of memory to ensure that the required gates are opened at the appropriate times. One way of achieving this is to have a circulating digital store which provides sequentially, at each time slot, the address of the gates to be opened. This store could take many forms: a

SWITCHING ELEMENTS 49

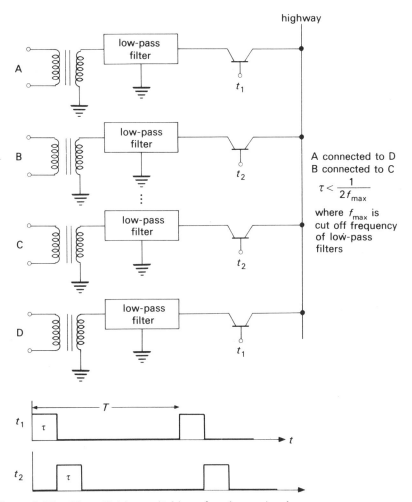

Figure 2.17 Time-division switching of analogue signals.

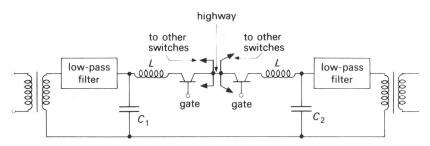

Figure 2.18 Principle of resonant transfer.

magnetic core store with one word per time slot containing the two addresses, a series of integrated circuit shift registers, or (in early systems) a delay line whose delay is equal to the period between sampling pulses. Systems based on this principle have been used successfully for small private telephone systems and also for some military switching systems which have to be mobile and therefore very compact.

The number of time slots in the above systems is limited by the fact that the sampling frequency cannot be less than 8 kHz. The number of slots in the 125 μs period between samples may be increased to 100 or more but τ/T becomes very small. Also with pulses of width 1 μs or less their propagation time and delay times through transistor circuits become significant. So the pulse duration must be less than the time slot in order to avoid the possibility of inter-channel interference caused by energy from one time slot moving into an adjacent time slot.

(b) Four-wire systems. If amplification is needed to make up for the transmission losses it is necessary to make the system four-wire as shown in Figure 2.19 [17].

The hybrid transformer splits the speech channel into GO and RETURN channels. Also it is designed to provide the necessary bandlimiting. Since the system is four-wire, separate time slots are needed for each direction of transmission.

(c) Switches for digital signals. If the signals to be switched are in digital form, either because they are binary digital data or because they are a digitalised form of an analogue signal (such as pulse code modulated speech), the switches may be made from logic elements. The requirements for crosspoints are then considerably relaxed compared with those for analogue crosspoints. This is because:

(i) The signal has to be identified only as 1 or 0. So the 'off' resistance does not have to be so high and since the signal may be regenerated, the finite 'on' resistance does not result in attenuation of the signal.
(ii) As long as the switch operates at a sufficiently high data rate there is no problem of bandwidth or power handling capacity.
(iii) If the signal is digital, alternative means will already have been made for signalling, so no d.c. path is necessary.
(iv) Two-way transmission is not needed.

If a simple digital switch is required for something like a telex switching system it is possible to provide this with a space-division switch using transistors or with a time-division switch using logic gates. However, because of the digital form of the information, there is a further possibility of using a storage medium. The principle of this is shown in Figure 2.20. Each inlet has a device for

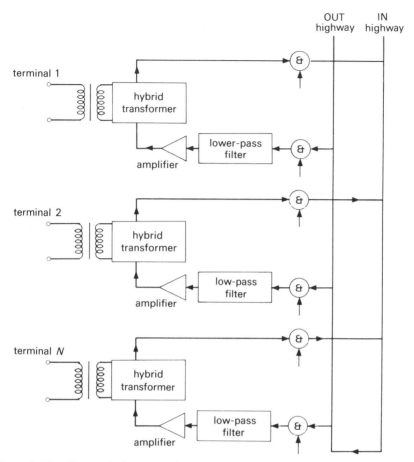

Figure 2.19 'Four-wire' p.a.m. t.d.m. system.

detecting and storing a complete character consisting of several bits. Once a character has been received, the common control places it in a specified location within the main store. On the outlets there are other devices which extract a character from a specified location within the main store, store it locally and transmit it serially to the outgoing line. The nature of the addressing mechanism used to access the main store means that any incoming character can be placed anywhere in this store and any outgoing character can be extracted from anywhere in the store. The store therefore acts as a very high capacity non-blocking switch.

This method introduces a delay equal to the transmission time of a single character, but this is usually acceptable. Some systems have been developed which introduce a delay of the transmission time of a single bit only [18].

52   SIGNALLING AND SWITCHING TECHNIQUES

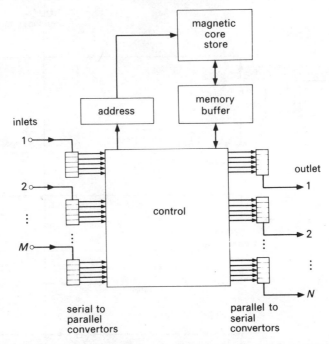

Figure 2.20   Use of storage medium to provide a digital switch. When an input serial-to-parallel converter has a complete character stored it signals the control which then stores that character in a specified store location associated with the terminal. When an output parallel-to-serial converter is empty, it signals the control which accesses a location in the store associated with the output terminal. This location contains the address of the location associated with the input terminal to which a connection is currently made. The contents of this location are then read out from the memory and transferred to the output parallel-to-serial converter.

Frequency division switching.   Although the use of frequency division switching has long been recognised as a possibility [19], it is only the advent of the communication satellite that has made its application economic. Such satellites can be used to provide communication between a large number of ground stations. The straightforward method of providing communication between many stations is to assign a specific set of carrier frequencies for each combination of routes. For low traffic this is an inefficient use of the capacity of the satellite. Frequency division switching can be used to increase the traffic capacity of the satellite by making all carrier frequencies available to all stations. In a system developed by COMSAT called SPADE [20] there are 800 carrier frequencies available and each ground station can transmit signals via any one of these carriers by means of a switched frequency synthesiser generating the carrier frequency. So, when a new call arises, a free carrier frequency is seized

SWITCHING ELEMENTS 53

and the identity of the carrier seized is signalled to the remote end via a common data channel. At the remote end of the transmission channel, the demodulator is tuned to this frequency.

References

1. Hills, M. T. and Evans, B. G. (1973) *Telecommunication Systems* (Chapter 2). Allen & Unwin.
2. Smith, S. F. (1974) *Telephony and Telegraphy A*. Oxford University Press.
3. *CCITT Orange Book* (1977) Technical features of push-button telephone sets, Vol. VI-1, Recommendation Q23.
4. *CCITT Orange Book*, (1977) Telegram retransmission system, Vol. II-B, Recommendation F31.
5. For example, Tymann, B. (1973) Bit-oriented control procedures for computer-to-computer communication systems, *Computer* (IEEE), pp. 27–31.
   and international specifications (for packet switching) in *CCITT Orange Book* (1977) Interface between data terminal equipment and data circuit-terminating equipment for terminals operating in the packet modes on public data networks, Vol. IX, Recommendation X25.
6. This technique is used in the new No. 1 and 1A ESS to provide 4-wire switching on a 2-wire network. See Haugk, G. (1976) The new peripheral system for No. 1 and No. 1A ESS, *International Switching Symposium*, paper 131–2.
7. For example, Rogers B. H. E. and Ridgeway (1965) Dry-reed relays – Post Office type 14 and type 15, *Post Office Elect. Eng. J.*, 58, pp. 46–9.
8. Feiner, A. (1964) The Ferreed, *Bell Syst. Tech. J.*, 43, pp. 1–14.
9. Gashler, R. J. *et al.* (1973) The Remreed network: a smaller, more reliable switch, *Bell Lab, Record*, 51, pp. 203–7.
10. For example, see Corner, A. C. (1966) The 5005 crossbar telephone exchange switching system, *Post Office Elect. Eng. J.*, 59, pp. 170–80.
11. Alexandersson, H. V. (1964) The L. M. Ericsson code switch – a new connection device, *Ericsson Review*, 41, pp. 80–87.
12. Takamura, M. *et al.* (1971) DEX-21 switches and relays, *Rev. Elect. Comm. Lab., Japan*, 19, pp. 344–49.
13. Vazquez, C. and Dufresnog G. (1969) Miniswitch. A miniature crosspoint switch, *Elect. Comm.*, 44, 4, pp, 288–92.
14. For example, see application of p.n.p.n. diodes in Stewart, A. (1973) TCS transmission aspects, *Elect. Comm.*, 48, pp. 375–89.
15. Cattermole, K. W. (1958) Efficiency and reciprocity in pulse amplitude modulated systems, *Proc. IEE*, 105B, pp. 449–452.
16. French, J. A. T. and Harding, D. J. (1959) An efficient electronic switch – The bothway gate, *Post Office Elect. Eng. J.*, 52, pp. 37–42.
17. Harris, L. R. S. *et al.* (1960) The Highgate Wood experimental electronic telephone exchange. General introduction, *Proc. IEE*, 107B, Supplement 20, pp. 70–80.
18. Mactaggart, D. (1976) A multiprocessor bit switch used as an international telex exchange – CMA-745, *International Switching Symposium*, paper 572–1.
19. Flowers, T. H. (1952) Electronic telephone exchanges, *Proc. IEE*, 99, pt. 1, pp. 181–193.
20. Puente, J. G. and Werth A. M. (1971) Demand-assigned service for the INTELSAT global network. *IEEE Spectrum*, 8, 1, pp. 59.

3
The design of economic switching centres

3.1 Introduction

The purpose of this chapter is to show the various ways in which the three techniques of resource sharing described in Chapter 1, may be applied to the Basic Switching System in order to reduce the total cost of a switching centre. Each method of resource sharing must be applied in such a way as to minimise the effect of faults on the system availability.

As explained in Chapter 1, the Basic Switching System of Figure 1.5 (page 13) is the most reliable system possible. A fault or malfunction within an individual control system obviously affects the terminal to which it is connected. If it is assumed that the speech switch can make a connection to only one other terminal at a time, then only one other line can be affected.

In such a system there is a certain amount of common equipment, notably power supplies and tone supplies. To protect against faults in these they must be duplicated. A failure of a power or tone supply is easy to detect and a simple change-over mechanism may be activated in the event of a failure. If the change-over mechanism itself goes wrong and change over occurs unnecessarily, no harm is done. Regular testing of the change-over mechanism should ensure that the probability of the mechanism failing to change over when needed is low. A complete system failure only occurs if a fault develops in the change-over mechanism in the interval between checks and, in that same interval, the main supply itself fails.

However, another type of fault that may cause supply failure is a short circuit within the system. In this case, operation of the change-over mechanism is of no use. This situation can be avoided by a system of fusing which ensures that only a limited number of control systems are fed from one set of fuses. Most fuses used in telecommunication systems are fitted with contacts which are closed

FUNCTIONAL SUBDIVISION 55

when the fuse blows: this closing of contacts can be used to send a signal to a manned control centre so that a repair procedure can be started.

So, provided these techniques are used for the common equipment a system like the Basic Switching System has a very high system availability: that is the probability of a complete system failure is very low, although individual users experience failures from time to time.

3.2 Functional subdivision

The simplest form of functional subdivision is that shown in Figure 1.9 (page 18), where the control unit is partitioned into a line unit and main control unit. The line unit must be able to:

— detect a *seize* signal from the terminal,
— select a free main control unit,
— connect itself and the speech wires to a main control unit,
— release itself when instructed by a main control unit.

In addition, for terminating calls, it must:
— give an indication to other main control units of whether it is busy or free,
— provide a connection to a calling main control unit for the duration of the call.

Note that the functions performed by the line unit are chosen according to a particular design approach, namely, for any particular call, there should be only one *centre of control*. That is to say, while a terminal is idle its centre of control is its line unit. Once a call request has been accepted, the centre of control is transferred to the selected main control unit. Later in the setting up of the call it is this control unit which sends the *seize* signal to the line unit of the called terminal, so the called line unit is subservient to the main control unit.

Alternative design approaches could lead to more autonomous control in the line units. For instance, even though a line unit is connected to a main control unit, it could continue to look for a *clear* signal, and release itself from the main control unit on detecting one. This particular partitioning of control would increase the complexity of the line unit. It would also decrease the flexibility of the main control unit because there are certain types of call where it is necessary to maintain a connection even though a *clear* signal has been received (e.g., a telephone call to an emergency service such as fire or police). So, in practice, the use of a single centre of control is usually the best approach.

Having decided upon the initial partitioning of the control unit, it is now possible to examine the requirement of the access switch arrangement used to associate a main control unit with one line unit. There are various possible arrangements:

(a) A hunting system.   This is the method shown in Figure 1.9 (page 18).

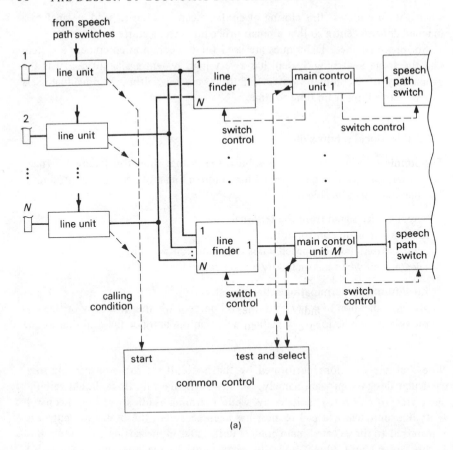

(a)

Each line unit has its own access switch and access switch control. This switch has access to a number of main control units and each control unit provides an indication of whether it is busy or free to each access switch. When a line unit detects a *seize* signal it

— hunts for a main control unit with a free indication;
— instructs the selected main control unit to busy itself to other line units;
— connects the control paths and message information paths from the line unit to main control unit;
— transfers the centre of control to the main selected control unit.

If more than one of the line units are hunting simultaneously for a free main control unit, it is possible for two line units to test and select the same main control unit. Special circuit techniques are used to minimise the probability of this happening; these are discussed in Chapter 9.

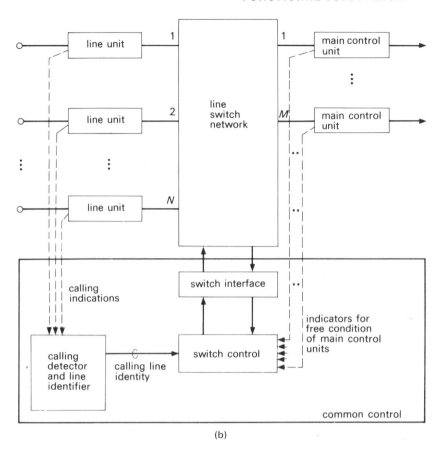

Figure 3.1 Examples of functional sub-division. (a) Line finders. (b) Line switch with common control.

(b) A finding system. The hunting system requires an access switch and a control for every line unit. An alternative is to associate the access switch and its switch control with the main control unit (Figure 3.1a). Because there are far fewer of these than line units the number of switch controllers (but not the total number of crosspoints) is reduced. This arrangement is called a *finding* system because the main control has to find a calling line unit rather than the calling line unit hunting for a free main control unit.

If the switches are electronic components it is feasible for each free main control unit continuously to scan all line units and look for one which is calling. In this case the system must be designed so that a line unit in a calling state is not connected to more than one main control unit. A straightforward way to achieve this is to introduce a common control which, for instance, permits scanning by only one main control unit at a time.

58   THE DESIGN OF ECONOMIC SWITCHING CENTRES

With electro-mechanical devices continuous scanning is impracticable and a common control is essential. A general form of common control is shown in Figure 3.1a. The common control has access to the busy/free indications of all the main control units; it also has an input from all the line units to indicate when one (or more) is in a calling state. On receipt of a calling signal, the common control selects a free main control unit and sends it a signal to instruct it to start searching for the calling line unit. Once a main control unit finds the calling line unit, it sends it an *accept* signal; this removes the calling condition from the input to the common control. The main control unit then signals to the common control that it has been successful.

The finding process takes a finite time and there may be occasions when further calls arise before the calling line has been dealt with. In the system described above the main control can find only one calling line at a time. In this case once one of the calling line units has been connected and the common control informed, there will still be a calling condition applied to the input of the common control. The common control will therefore assign another free main control unit to find the next calling line unit.

If new calls arise while there is no free main control unit, these calls are ignored by the common control until a main control does become free. The common control therefore imposes a queueing discipline, though not necessarily a strict one: the order in which calls are connected may not be their order of arrival. The order of serving depends upon the design of the mechanism which looks for the calling conditions.

(c) Common switch network.   A third access method is the use of a common switch network, as shown in Figure 3.1b. The main difference between this and the previous technique is that the common control is now more powerful. Between the common control and the line unit is an identification system which provides a coded indication of the actual line unit making a call. (This is instead of a common line in the hunting system which simply indicates that one, or more, of the line units is calling.) The identification system is constructed so that it identifies line units one at a time. If two or more calls arise within a short time (or while no free main control unit is available) they will be identified sequentially. In an electronic system this may be achieved by scanning the line units sequentially and stopping the scanner if a calling condition is found. Electro-mechanical systems require more complex arrangements, some of which will be described in Chapter 9.

Because the common control for this third access method is provided with the identity of a calling line unit and because it also has access to the busy/free indications of the main control units, it can therefore make a decision about which main control to use and can instruct the switch network to set up a suitable path between the calling line unit and selected main control unit.

One of the advantages of this type of system which will become more apparent in Chapter 7 is that the switch network may consist of several stages of

FUNCTIONAL SUBDIVISION   59

switches. In this case there may sometimes be free main control units to which no free path is available from the calling line unit. This type of common control is able to assess the path availability and select a free main control unit to which there are suitable paths. The practical effect of such *conditional selection*, as it is called, is greater efficiency in the use of the switch network.

The effects of malfunctions in functional subdivision systems.   We shall now consider the effects of malfunctions on the system availability for these three techniques. A malfunction occurring in the line unit or its associated circuitry will affect the associated terminal, as described earlier. In all three cases the complexity of equipment associated with an individual terminal is reduced as compared with the Basic Switching System, so the probability of a malfunction ought to be lower.

In the hunting system, a malfunction occurring in a main control unit affects the line unit(s) to which it is currently connected. The effect on overall system availability is minimised if:

- the sub-system is designed such that, even for most expected malfunctions, it will still respond to a *clear* signal and release the connected line unit;
- the access switch mechanism is arranged to pick a free main control unit at random, it is then likely that a terminal making a second attempt will use a different main control unit;
- a regular check of the system is made to discover and rectify faulty main control units and thus prevent the accumulation of faulty sub-systems.

With these techniques, the system as a whole will work effectively even though a number of calls may require repeat attempts before they are connected.

Both the finding arrangement and the common switch network arrangement introduce further problems because they need a common control. A malfunction of this common control could affect the whole system. Two alternative ways to overcome this are:

- building the common control with enough internal redundancy for its probability of failure to be acceptably low. (Techniques for provision of redundancy are discussed later in this chapter.)
- limiting the number of line units which depend upon a common control. In this case a single malfunction affects only a proportion of terminals. The applicability of this method depends on the system availability required, the expected fault rate, and the fault detection and repair time.

In addition, the individual main control units must have a low probability of indicating that they are free when they are faulty or when they are in use in order to prevent double connections. This implies that the free condition should be indicated positively and its absence should mean non-availability through being busy or faulty.

Extension of functional subdivision technique. The principle of functional subdivision may be used at many levels. In any switching system there are seven basic functions required:

(a) *Call detection* — detecting the *seize* signal from a terminal (or a trunk).
(b) *Routing signal reception* — this is generally a sequence of signals which have to be detected and stored. In switching systems these functions are collectively known as the *register* function.
(c) *Translation and test* — determining from the received *routing* signal either the identity of the required terminal or a suitable outgoing trunk group. The status of the required terminal or trunk route (free or busy) must then be tested. If the terminal or trunk group is busy, a suitable signal must be returned to the originating terminal (or switching machine). If there is more than one free circuit in a required trunk group, a particular one must be selected.
(d) *Path search and set-up* — finding and connecting a suitable path through the switch network between the calling and called units.
(e) *Sending* — transmitting a *seize* signal to the required terminal or, if the call is for another switching centre, transmitting call progress signals to the calling terminal and repeating *routing* signals to the trunk.
(f) *Supervision* — monitoring the call, once a path has been successfully set-up in order to detect answer and clear down conditions and, in some cases, bring about recording of billing information.
(g) *Clear down* — clearing the path, and any associated control system, once the supervisory function has determined that the path is no longer needed.

These functions fall into three classes:

— per terminal functions (needed all the time),
— per connection functions,
— per set-up functions (including clear down).

It is possible to split up the set-up function as well as shown in Figure 3.2. This shows that the main control unit is used for the whole connection but is connected to a sub-system called a *register* for the duration of the call set-up. The translation function is only required for part of the set-up time, so this is removed from the register to a separate system.

As will be seen later, there are a large number of ways of splitting up the control unit into smaller sub-systems and then connecting the resulting sub-systems together for a particular call.

A note on terminology. Telephone switching systems have been developing for nearly a century and a wide variety of names has arisen for sub-systems which provide basically similar functions. For instance, in the system of Figure 3.2 the

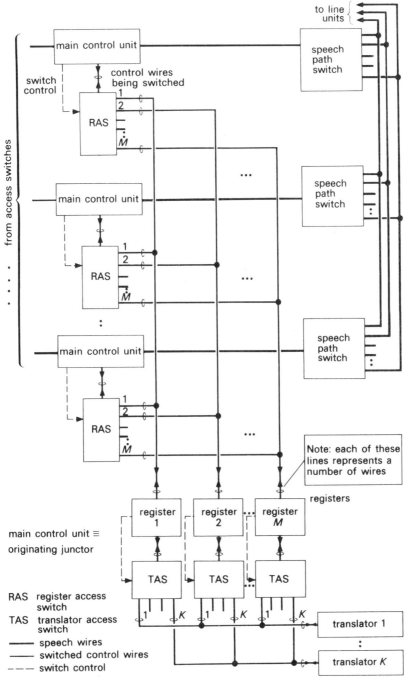

Figure 3.2 Example of the splitting of a control circuit into functional sub-systems.

sub-system referred to here as the main control unit is called, by various manufacturers:

> transmission bridge,
> trunk relay set,
> intra-office trunk circuit,
> supervisory unit,
> junctor.

In this book the term *junctor* will be used. This term was originally used in crossbar exchange systems and describes any circuit between two switching sub-networks; this circuit may be anything from a pair of wires to a complex control unit.

In the example of Figure 3.2, the main control unit is called an *originating junctor* because it is connected to the originating line unit. If, as in this example, the junctor can be connected directly to the terminating line unit, a more accurate description is *local junctor*, that is the junctor used for local calls. Other examples will be found later in this chapter.

In a software-based data or message switching system, the same principle of functional subdivision may be used. | However, in this case, it is groups of software sub-routines which provide the functional subdivision.

3.3 Common switch networks and their control

A considerable reduction of the total number of crosspoints in a switch network may be achieved if a series of one-input, many-output switches are replaced by a many-input, many-output network. This will be discussed in detail in Chapter 7, where it will be shown that the price paid for this is the possibility of *internal blocking*. This means that even though a particular terminal (or device such as a register) is free, it might not be connected because, at the required time, no suitable path through the network is available.

For a switch network the control which operates and releases individual mechanisms is needed for only a small proportion of the time the switch is in use. So, further cost savings are possible if the switch control mechanism is shared between a number of individual switches. With many switching mechanisms (such as crossbar or co-ordinate arrays of reed relays) such sharing is basic to the switch design. The use of a shared switch network, therefore, generally implies the use of a common control for that network. One example has already been given in Figure 3.1b.

A further application of this method of resource sharing is shown by the progression from Figure 3.2 to Figure 3.3a. In this case all the switches performing a particular function are grouped together into a common block. The building block is shown in more detail in Figure 3.3b. There are, in fact, two versions of this building block:

COMMON SWITCH NETWORKS AND THEIR CONTROL 63

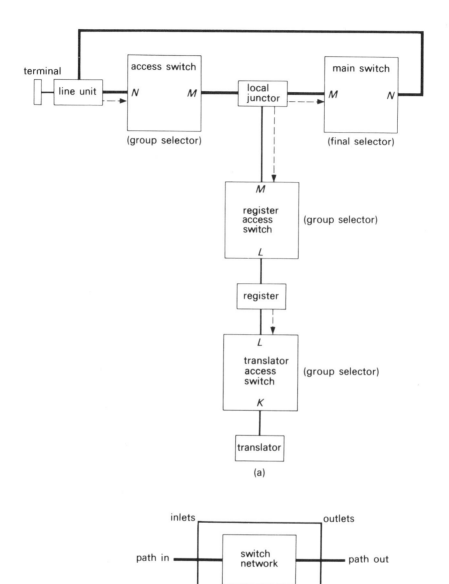

Figure 3.3 Use of common switch networks. (a) Use of common switch network. (b) Structure of switch network and control.

64   THE DESIGN OF ECONOMIC SWITCHING CENTRES

- *a group selector* which, on receipt of a command from a unit connected to one of its inlets, finds a free unit connected to one of its outlets and a suitable path to it;
- *a final selector* which, on command from a unit connected to an inlet will find a path to a *particular* outlet if the unit connected to that outlet is free.

As may be seen in the simple example of Figure 3.3, all but one of the switch networks are group selectors since, apart from the main speech switch, any free unit on the other side of the switch can be selected. The main speech switch must be a final selector since (in general) only a particular terminal will be called. It will be seen later that the concept of a group selector can be extended to switch networks which have outlets connected to units in a number of groups, for example, one group for each trunk route. In this case, the function of a group selector is to connect a given inlet to any free unit within a particular group.

Larger switch networks.   The network shown in Figure 3.3 is very simple. This kind of network might be used in a small switching centre (say up to 200 terminals). In practice, larger centres are often needed and they must have the ability to grow easily. For example, a telephone switching centre may have only 5000 terminals when it is installed, but it may grow to over 20 000 terminals in a few years. To achieve economy and the ability to grow, and also to maintain high system availability, all telephone switching centres of over a few hundred terminals have switch networks arranged in blocks as shown in Figure 3.4a. Here, as a specific example, a 1000-line system is constructed from ten 100 x 10 access switches giving access to a total of 100 junctors. These 100 junctors have access to a group of 10 registers. Each group of 10 junctors have access to its own group selector which in turn provides 4 circuits to each of 10 final selectors. Each final selector gives access back to 100 lines. This network is more simply drawn in Figure 3.4b, where the details of interconnections are removed.

3.4   Control of switching systems

The previous section explained some possible organisations of networks, we now examine how they can be controlled. This will be done with reference to Figure 3.4b. With several final selector networks available from a particular local junctor it is necessary to arrange for the transfer of control information from the register to the correct network. There are a number of techniques to achieve this, ranging from those with distributed control to those with common control.

'Step-by-step' or distributed control.   With this technique the register instructs the group selector of the local junctor to which it is connected, to set up a path to a free circuit which provides access to the required final selector network. There is then a path established from the register to the required final selector,

CONTROL OF SWITCHING SYSTEMS 65

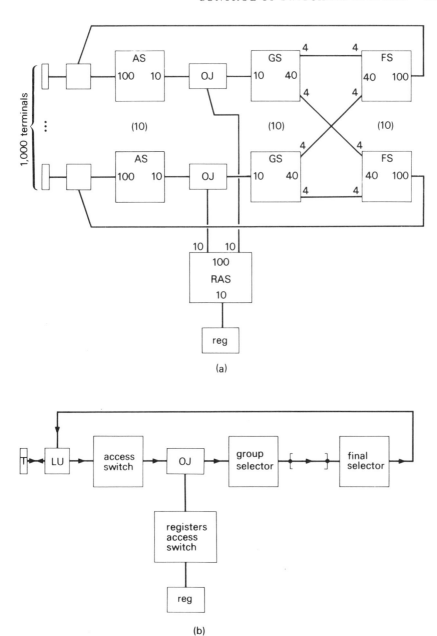

Figure 3.4 Extension to large systems. (a) Example of 1000-line switch network. (b) Simplified schematic of (a).

via the local junctor and associated group selector. The register can next instruct that particular final selector network to establish a connection from this path to the required line unit. If the line unit is free it is seized, and the path is completed between calling and called line unit. However, if it is busy, either the final selector network itself returns a busy signal to the calling line, or a signal is returned to the originating junctor (or perhaps to the originating line unit), instructing it to send a busy signal to the terminal.

Common control. With this technique common control of the switch network is used. Typically the action of the common control is as follows (Fig. 3.5):
— Once the register has received the *routing* signal it sends a *seize* to the common control.

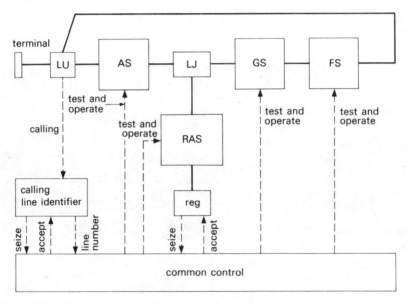

Figure 3.5 Use of single common control.

— When the common control is ready to serve the calling register it returns an *accept* signal and the register then forwards a copy of the *routing* signal.
— The common control tests the required line unit to determine whether it is busy or free. If the line unit is busy, the common control arranges for the local junctor to transmit a *busy* signal to the calling line.
— If the called line unit is free, the common control accesses the controllers of the associated group selector network and the required final selector network and finds and connects the path.

CONTROL OF SWITCHING SYSTEMS 67

This type of common control is often called a *marker*.

The common control technique may be extended even further to provide the control functions required for the initial path set-up from the calling line unit to the originating junctor and to the register, as shown in Figure 3.5.

Availability aspects of the system organisation

Step-by-step. The step-by-step control system has a high tolerance to individual faults. Each network controller is independent of the others, so a malfunction in a controller or its switch network affects only a proportion of calls. The tolerance to malfunctions can be made even better if:

— the access switch is provided by a hunting arrangement (so it has no dependence on a common control);
— the registers are either provided with separate controllers or, if they do depend upon common control, are split into a number of groups, each with its own common control;
— each originating line unit has access to a number of different group selector networks so that a malfunction in one affects only a proportion of calls;
— a number of independently controlled final selector networks have access to the same called line unit.

These four principles are shown diagrammatically in Figure 3.6. This shows a single line unit with access to a number of originating junctors, each with its separately controlled registers, register access switch and group selector networks. The group selectors have access to separately controlled final selector networks which have independent access to the same called line units. Provided that the originating line unit chooses an originating junctor at random and that the group selector networks choose a final selector network at random, then

— a malfunction in the line unit area will affect only one (or two) terminals;
— a malfunction in any other part of the system will (in general) affect only the traffic carrying capacity of the system and cause some repeat attempts by a user.

These techniques are the bases of the very high system availabilities that can be achieved with distributed control. They are used, in particular, by the step-by-step systems such as Strowger switching systems.

Common control. A distributed control system such as described above is one with an inherently high system availability. However, a common control system has an inherently very low system availability because a single fault in the common control area could make the whole system inoperative. The probability of this happening can be reduced by using a number of controllers as shown in

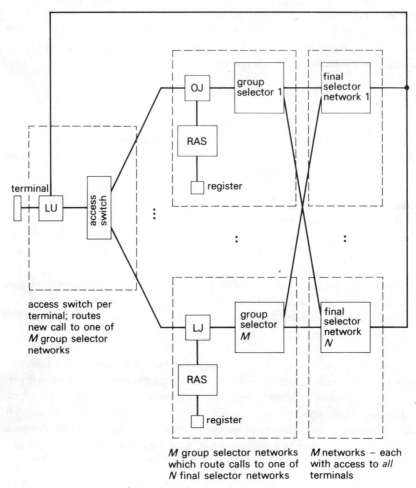

Figure 3.6 Principle of distributed control.

Figure 3.7. In this case each line switch network can detect a calling condition and seize one of a number of controllers to perform the call set-up. A high system availability can be achieved if:

— the devices which can call in a controller choose the one they are to use at random;
— the devices can detect gross malfunctions of the controllers. For instance, if a controller is not connected or not released within predetermined times the calling device can make an automatic second attempt via another controller;
— the controllers are designed such that they have no stored information relating to particular calls, so different controllers may be used for the initial and final path set-ups. This is called *memoryless* marker arrangement.

CONTROL OF SWITCHING SYSTEMS 69

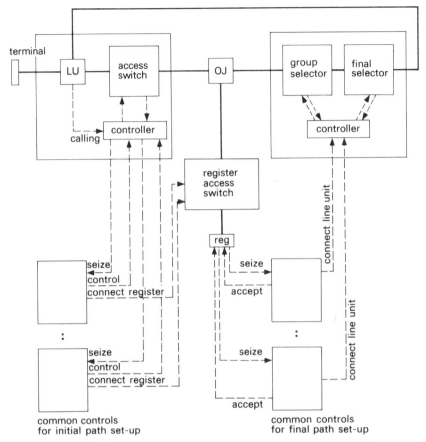

Figure 3.7 Use of multiple common controllers to improve system availability.

The system above is, in fact, the basis of the majority of crossbar switching systems in use today.

Redundancy techniques. When a system has to be dependent upon, effectively, a single common control, some form of redundancy must be provided in order to achieve the required level of availability. This redundancy may be provided in several ways:

Worker/standby. In a worker/standby arrangement two sets of a common control are provided with a change-over mechanism (Fig. 3.8a). Change over must occur when the worker system goes wrong. The problem is to decide when a change over is needed. Some form of malfunction detection is needed to do that. The simplest method is to use a watchdog timer which initiates change over if

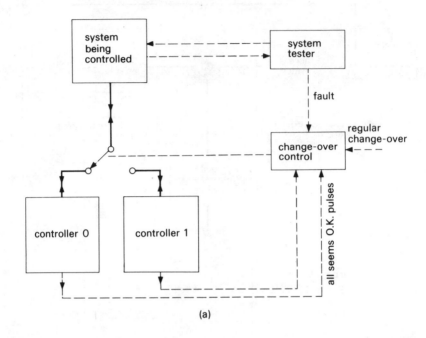

(a)

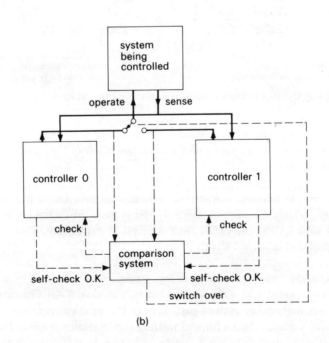

(b)

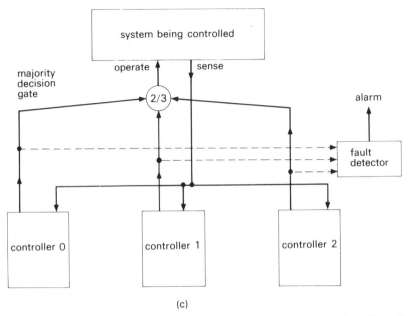

Figure 3.8 Methods of provision of redundancy. (a) Worker/standby. (b) Duplication. (c) Triplication.

the common control does not complete a sequence of operations within a predetermined time. More complicated malfunction detectors may be used to send test calls automatically through the system and look for the correct responses. If the correct responses fail to come within a predetermined period, a malfunction is assumed to have occurred; a change over is initiated and an alarm given. A regular change over is necessary to ensure that the standby does not develop any undetected fault and (for electro-mechanical controllers) to equalise the wear on the systems. It is also necessary to check regularly that the fault detector works so that any malfunction in this does not go unnoticed. In general, a malfunction in the fault detector producing a false alarm will not cause problems if the standby is fault free.

Duplication. Worker/standby systems are adequate if there is no accumulated information stored in the common control. When there is accumulated information a change over to the replacement control will start without this information. This implies that all calls in progress have to be forcibly released so that the information in the new standby control can be built up from a known starting point. To overcome this problem it is possible to have a duplicated system in which both controls are permanently connected to the inputs from the inter-working sub-systems (Fig. 3.8b). Both controls should then have the

72 THE DESIGN OF ECONOMIC SWITCHING CENTRES

correct memory contents. However, only the working control is able to issue output commands to the controlled sub-systems.

The use of duplicated controls permits a convenient fault detection method: the outputs from each control can be continuously compared. They should normally be the same, although only one set is used to activate the controlled sub-systems. When a discrepancy is detected it will be known that a fault has occurred, but it will not be known which control is in error. If the controls are processor-like devices, a discrepancy may be used to trigger a self-check programme run independently within each control. The processor which completes its checks first will then be placed in control.

Triplication. It is possible to use three controls with a means of comparing their outputs (Fig. 3.8c). In this case the 'odd man out' is the faulty controller. In most commercial switching systems the cost of triplication is usually too high to make this a practical solution.

3.5 Alternative organisations of switching networks

So far only the structure of Figure 3.4 has been discussed. There are, in fact, a vast number of alternatives to this structure, using different ways of partitioning the basic control system functions and different arrangements of switch networks.

Line and trunk switch networks. In all types of switching equipment, with the exception of Strowger (which is discussed in Ch. 10), it is possible to use a

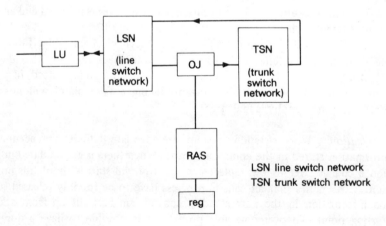

Figure 3.9 Use of (line switch networks) LSN and (trunk switch networks) TSN. (Arrows show traffic flow.)

ALTERNATIVE ORGANISATIONS OF SWITCHING NETWORKS 73

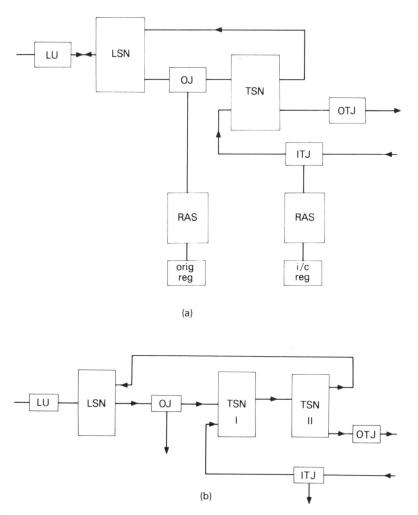

Figure 3.10 Provision of transit switching and improved access to outgoing trunks. (a) with single TSN. (b) with additional group selector stage.

switching network for both directions of traffic. In other words, the same physical network may be used as an access switch and as a final selector. This leads to the network structure of Figure 3.9. Note that there are actually many switch networks of each type.

In the systems of the form of Figure 3.9, the usual terminology is to call the switch network connected to the terminals the *Line Switch Network* (LSN) and the internal network the *Trunk Switch Network* (TSN).

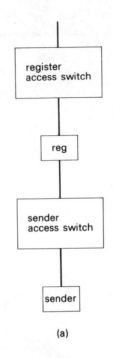

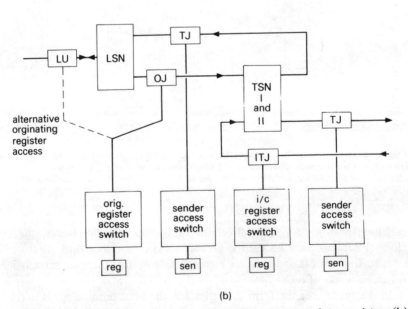

Figure 3.11 Connection of senders. (a) Sender connected to register. (b) Direct connection of registers and senders to TJ and OTJ.

ALTERNATIVE ORGANISATIONS OF SWITCHING NETWORKS 75

This structure may be extended to cater for connections to other switching centres. Figure 3.10a shows one possibility. If the register decides that a call is for another switching centre, it instructs the control to connect the originating junctor to a free circuit within the required trunk group to that centre. The register then forwards the relevant routing information to that centre. The register in this case includes the *sender* function.

A call incoming from another switching centre needs access to a register. When this register has determined the destination of the call, it instructs the control to set up a path to the required terminal.

Note that this configuration appears to offer the possibility of transit calls since an incoming trunk may be connected to an outgoing trunk via the TSN. However, this is of limited use as in general there will be many TSNs. As shown, one incoming trunk has access to only one TSN and therefore only a sub-set of the possible outgoing circuits. (It does, of course, have access to all the LSNs.) To cater for transit calls an additional group selector stage is required so that an incoming trunk call may be given access to all TSNs (Fig. 3.10b).

Alternative allocation of junctor functions. The type of the interface provided by a junctor for local lines is different from that for trunk lines. For a local line it must provide a d.c. feed, transmission of audible tones, transmission of ringing currents, and so on. The interfaces to trunk circuits have many forms, depending particularly upon the line signalling technique used. Since there is such a variety of interfaces, the junctors shown in Figure 3.10a can become complex; they have to connect to a variety of lines on the other side of the switches. For this reason the functions of the junctors are often split into two devices such that each trunk or line unit is connected to its own type of junctor, and these junctors provide a common standard switching-side interface to each other. Figure 3.11a shows an example of this.

The functions of a local junctor are then split between:

— *an originating junctor* (OJ) to interface a calling terminal (via a line unit);
— *terminating junctor* (TJ) to interface a called terminal (also via a line unit).

The trunk junctor is split into:

— *an incoming trunk junctor* (ITJ) which interfaces to the trunk circuit on one side and to a terminating junctor on the other;
— *an outgoing trunk junctor* (OTJ) which provides access to the outgoing trunk from either an originating junctor or, for a transit call, from an incoming trunk junctor.

When a trunk circuit is operated 'both-way' (in terms of traffic) a *both-way trunk junctor* is required; this combines the functions of ITJ and OTJ.

76   THE DESIGN OF ECONOMIC SWITCHING CENTRES

Use of senders.   Other options of functional subdivision relate to the register function. As described so far, this includes a sending function. The type of sending required depends on the outgoing route. Further economies are therefore often possible by providing separate senders for each type of signalling system, rather than a single general purpose sender within the register. There are many different means of connecting the senders into the network. Figure 3.11a shows one possibility: they may be connected when needed to the register. Alternatively, they may be connected directly to the terminating junctor or outgoing trunk junctor, as shown in Figure 3.11b. This latter arrangement has the advantage that the junctor requires access only to the type of sender relevant to the route to which it is connected. However, it implies that a mechanism must be provided to transfer control information from the register which receives the routing information to the sender connected to the outgoing circuit.

Figure 3.11b also shows an alternative position for connecting the originating register.

Access switch sharing.   There is one more level of sharing possible with the use of common switch networks and this is to use the Line and Trunk Switch Networks as register and sender access switches. This leads to a common switching network used for all purposes. Two variants are shown in Figure 3.12.

Figure 3.12a shows an extreme example of this in which a single switch network can interconnect any pair of inlets. All junctors, registers, senders, and so on, are connected to this switch. The type of system is often referred to as a serial trunking system. Another alternative is shown in Figure 3.12b. In both cases their structure imposes constraints upon the control system organisation. These are discussed later.

It is difficult to generalise about the relative advantages and drawbacks of all these different arrangements. The choice of arrangement depends upon the size of the centre and its traffic capacity, the technology used, and, most importantly, upon the overall control philosophy.

3.6   Time-shared decision making

Types of control system.   The third technique of resource sharing is that of time-sharing the decision making portion of a control system. A control system consists of a control circuit connected between an input and an output interface, as shown in Figure 3.13. The input interface converts signalling information to a form suitable for input to the control circuit. The output interface converts signals from the control circuit into operations on the switch network and signalling information sent to the terminals and to other control systems.

The inputs and outputs of a control circuit are always discrete values, that is they can only take one of a limited number of values. The most common values are the binary values of '0' or '1'. These may be represented in many different ways depending upon the technology used. The information itself may be

TIME-SHARED DECISION MAKING 77

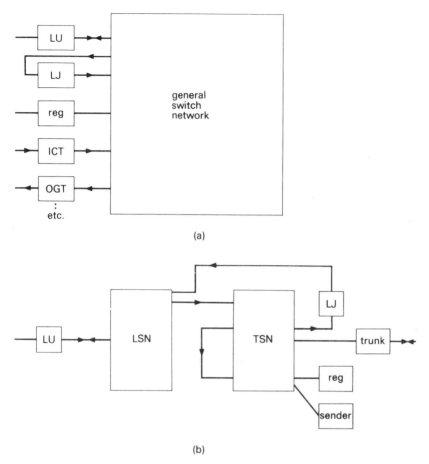

Figure 3.12 Examples of use of LSN and TSN for access functions. (a) General switch network (connects any pair of inlets). (b) Alternative to (a).

generated or transmitted in an electrical form such as a high or low resistance, or the frequency of a tone. The function of the interface is to convert the information from one electrical form to another.

The simplest type of control circuit is one where the output is at any time a function of only its inputs at that time. This is called a *combinational circuit* because the outputs are always given by combinations of the input conditions. An example of a combinational circuit is a decoder with four inputs, which correspond to a number in a binary code, and ten outputs, each of which is activated by a particular input combination.

In most control systems the signals received from a terminal or another control system have a meaning which depends upon signals which have been sent previously. The output from the control circuit at any given time must be a

78 THE DESIGN OF ECONOMIC SWITCHING CENTRES

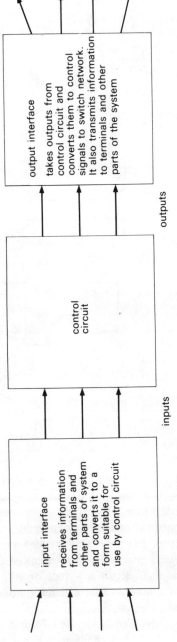

Figure 3.13 General control system.

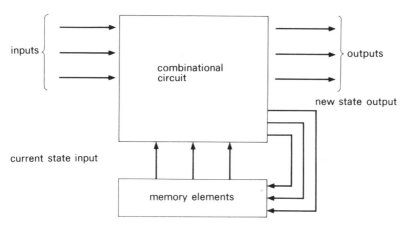

Figure 3.14 General form of a sequential machine (asynchronous working).

function not only of its input at that time but also of previous inputs which have since been removed. This implies that the control circuit must have some memory of the previous inputs. Control circuits with this property are called *sequential circuits*.

A general way to regard a sequential circuit is to separate it into a combinational part plus memory elements, as shown in Figure 3.14. If there are N memory elements, each of which can have one of two states, the memory will have 2^N possible combinations of stored information. The particular value stored in this memory is called the *internal state* of the sequential circuit. The output of the combinational circuit is a function of the input from the interface and the internal state.\ The output from the combinational circuit also modifies the internal state in order to make the outputs of the control system a function of previous inputs. The new state will be a function of the inputs at a given time and the current state.

There is a hidden timing problem in Figure 3.14 because, as the new state is stored in the memory elements, it will change the input to the combinational circuit. If the individual memory elements do not change simultaneously with the storing of the new state, intermediate states may be momentarily presented to the combinational circuit. This could give an erroneous new state and thereby cause the sequential circuit to operate incorrectly. In sequential circuit design this is referred to as a *race hazard*. Many design techniques have been developed to avoid the occurrence of race hazards. One simple technique is to ensure that only one memory element is changed whenever the state is changed, but more sophisticated techniques are available.

Timing problems are avoided if the sequential circuit is operated *synchronously*. This means that it can change its state only at discrete intervals of time. A synchronous sequential circuit is controlled by a regular series of clock pulses. One form of this type of circuit is shown in Figure 3.15; this uses two clock

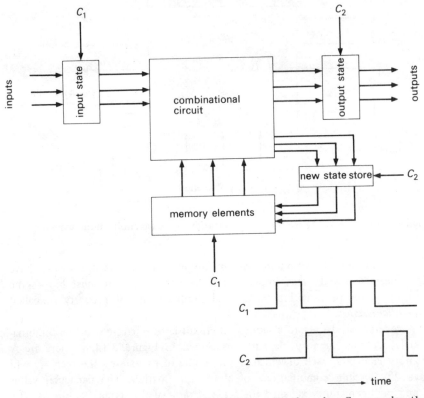

Figure 3.15 Synchronous sequential machine. Clock pulse C_1 samples the input conditions and stores them temporarily in the input store. It also replaces the state in the memory elements from the new state store obtained from previous cycle. Clock pulse C_2 will sample the output and new state from the combinational circuit and store them.

pulses. The first clock pulse samples the current state of the inputs and stores them in the input store. This clock pulse also changes the main memory elements to the new state from the previous clock cycle. The second clock pulse stores the output from the combinational circuit in an output store and a new-state store.

Time-division of control circuits. A switching system has to handle a number of concurrent calls which can arise at random instants. The simplest control system would be one which dealt with a call when it arose and ignored all others until the path had been set-up. This is a practicable arrangement where the calling rate is low or when the complete set-up time is so short that the probability of a new call being delayed is below an acceptable limit. If these conditions are not met it is necessary to provide a control system which can control more than one call set-up at once.

When there is a number of separate control sub-systems, each capable of performing the same set of functions, this is a space-division of control, as shown in Figure 3.16a. A control circuit built from electronic components has a potential speed of operation far faster than is needed for most applications. For instance, in a control system designed to respond to dial pulses the shortest event is a 33 ms make pulse between the dial breaks. Therefore, a sequential machine designed on a synchronous basis (that is one that can change its state only at discrete time intervals) need sample the input interface only every 10 ms (say) in order to receive these dial pulses adequately. An electronic sequential machine could change its internal state and give a new output in less than 1 μs.

The result of such a sample is applied to the combinational circuit together with the current internal state. The combinational circuit generates the new output and new internal state. The new output is transferred to the output interface and the new internal state replaces the current state.

The combinational circuit 'decides' on the new output and the new state. If this decision can be performed rapidly compared with the sampling rate it is possible to time-share the combinational circuit over several control sub-systems. This can be done as shown in Figure 3.16b. A switch is needed to select an input and output interface and connect them momentarily to the combinational circuit. At the same time the internal state corresponding to the selected control circuit is retrieved from a memory block. The combinational circuit then generates the new state and new output. The new output is transferred to the output interface, where it is stored and later used to produce the required actions. The new state is stored in the memory block.

The next input and output interfaces are connected to the combinational circuit and the output and internal state up-dated if the inputs have changed since it was last sampled. As long as the inputs are sampled within 10 ms, all control sub-systems will operate satisfactorily.

The above is an example of the concept of time-sharing of decision-making logic. Each sub-system requires its own memory for its internal state, and also memory for any outputs which have to be maintained between connections to the decision-making logic. It is just the equipment to change the internal state and outputs which can be time-shared. This time-sharing will save the cost of $N - 1$ combinational circuits but it introduces the extra expense of providing access circuits and storage in the output interfaces. With some memory technologies there is a further cost saving since memory in bulk may cost less per bit than memory provided in small chunks.

The introduction of time-sharing obviously involves common control and therefore, as discussed in previous sections, either redundancy or limited dependence must be used if high system availabilities are required. So, time-division is attractive only if the savings are not outweighed by the cost of the reliable provision of access switches, common control and more complex output interfaces. With the increasing cost effectiveness of modern electronics it is becoming less attractive to time-divide any but the most complex control circuits.

82 THE DESIGN OF ECONOMIC SWITCHING CENTRES

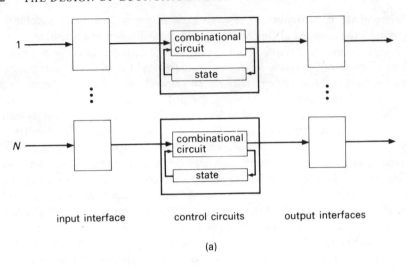

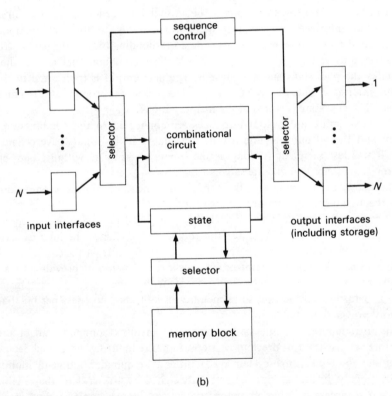

Figure 3.16 Different techniques for making N control systems. (a) Space-divided control systems. (b) Time-divided control systems.

3.7 Stored program control

The combinational circuit of the previous section can be realised in a more general-purpose control device as a look-up table containing entries indexed by the new input and old state. The entries give the new output and the new state. One realisation of a look-up table of this form is a read-only memory. As an example, if there are 2 inputs and 2 outputs and the number of internal states is 16 (2^4) the total number of inputs to the combinational circuit is 6. Therefore a read-only memory with 64 words could provide 1 word for every combination of input and state. Each word must contain 6 bits (2 for the output and 4 for the next state). The size of memory required increases rapidly with the number of inputs and states, but this technique (with some elaboration) is usable for a variety of simple control systems, especially if they are not likely to need modification during their operational life.

Examination of the contents of such look-up tables frequently shows a pattern. If there is a pattern this can be expressed as a set of rules to give the new output and state as a function of the input and old state. In these cases it is attractive to consider the use of a computer-like device to perform these rules, rather than a straightforward look-up table arrangement using special-purpose logic circuits. A computer can be programmed to test the conditions of the inputs and old states and decide on new outputs and states.

This principle is the basis for the modern technique of *stored program control* (or s.p.c., for short) of switching systems: the computer controller provides a time-shared decision maker. The decisions are expressed as programs which can be rewritten to modify or extend the functions of a control system. The use of computer technology also permits the application of memory elements for the state memory and, as will be seen later, for the switching of control information between the various control systems embodied in the one computer system.

It should be noted that replacing a look-up table by a program does not replace the combinational nature of the operation, even though the computer proceeds through several internal states while producing the result. This situation is similar to the two well-known computer techniques of deriving the sine of an angle: either the program stores all possible results against the angle or it computes the angle using an algorithm. In both cases the result is the same. The look-up method usually requires more storage but takes less time to deliver the result. In practice a method somewhere between the two extremes is used, depending upon the economic balance between cost of computing time and cost of storage. In using computers to control switching systems, exactly the same considerations apply.

An example of s.p.c.: register translator system. A simple example of s.p.c., shown in Figure 3.17, is the use of a computer to perform the control functions of the registers in a switching centre. Here a group of N registers have an associated single computer. This is connected to N input/output interfaces on

84   THE DESIGN OF ECONOMIC SWITCHING CENTRES

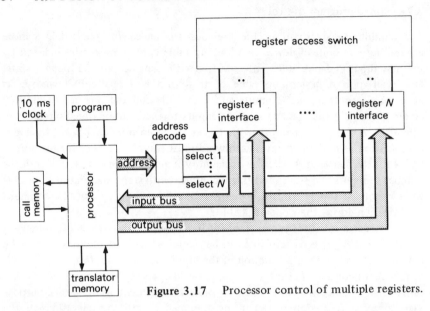

Figure 3.17   Processor control of multiple registers.

the register side of the register access switch. A particular interface may be connected to the computer input and output buses by loading the address of the interface into the address buffer.

The computer is programmed so that its memory is divided into two areas:

— program memory (for the sequence of instructions or look-up tables);
— call memory (for the state of each register).

In addition it may have a third area:

— translation memory.

The translation memory is used to relate actual dialled codes to specific actions, such as the setting up of either an internal call or a trunk call. The information in this memory could be regarded as part of the program, but it is more convenient to treat it separately. The program then remains a general algorithm which may be applied to any similar system. The specific (and possibly changing) aspects of particular systems are kept separate.

Each register interface consists of one or more inputs which have to be scanned to regular intervals. The computer accesses these scan points by loading the number of the required register into the address buffer. This address is decoded and the state of the associated scan points is returned. The outputs of each interface are treated similarly. The operate or release command is stored in a bistable memory element for each operate point.

The call memory is divided into a fixed block of words for each register. Typically 16 words of 16 bits might be used. The first word would indicate the overall state of the associated register and the rest would be used for assembling the dialled digits, storing them, and storing their translation.

The call memory may be regarded as being addressed by two indices. One index relates to the register currently being accessed by the computer. The other index is provided by the program and indicates the position of a word to be used relative to the start of the current block. This makes it possible to use the same program for all the registers.

In operation, the computer is triggered every 10 ms by an external clock pulse. An overall control program loads the address of the first register into the input/output buffer and sets one of the call memory indices to point to the block of memory associated with the first register. The overall control program then hands over to the register program which decides whether any action is required for the current state and current input. The actions include:

— sending instructions to operate or release relays on the output interface;
— modifying the stored information in the data block;
— modifying the stored state so that a different sequence of operations is performed on the next scan.

When the register program has completed its actions the overall control program modifies the addresses on the input/output buffer and the call memory index to address the next register, and so on. Once all registers have been dealt with, the computer waits until the next clock pulse before restarting a new scan.

Chapter 11 deals with the design of complete s.p.c. systems.

4
Traffic theory

4.1 Basic equations

The use of resource sharing implies that there will be occasions when a request for a connection cannot be served immediately and this request must either wait or be abandoned. The concepts of traffic and of grades of service were introduced in Chapter 1. This chapter shows how some simple theoretical results may be applied in the choice of a suitable level of resource provision in order to meet required grades of service. More exact theory is often needed for detailed design and this is covered in a number of references.

The basic theory depends upon the statistical parameters of user behaviour. The statistical parameters of a group of users varies with time, but for the purposes of this basic theory they can be taken as constant over a period such as the busy hour. The statistical analysis can then be based on conditions of *statistical equilibrium*. This implies that the probability per unit time of a new call request arising is equal to the probability per unit time of an existing call terminating. Hence, on average, the number of connections remains more or less constant.

If the average number of calls to and from a terminal during a period T seconds is n and the average holding time is h seconds, the average occupancy of the terminal is given by:

$$\rho = \frac{nh}{T} \qquad (4.1)$$

Assuming that calls originate at random from all the terminals, ρ represents the probability of a terminal being busy. According to standard statistical theory, for N independent terminals the probability of k of them being busy is given by the binomial (or Bernoulli) distribution

$$P(k) = \binom{N}{k} \rho^k (1-\rho)^{N-k} \qquad (4.2)$$

where

$$\binom{N}{k} = \frac{N!}{k!(N-k)!}$$

is the binomial coefficient.

As an example, consider the case of 10 terminals each generating 0·5 erlang. Figure 4.1 shows the probability of there being between 0 and 10 of the terminals busy. Figure 4.1 also shows the probability distribution for the same total traffic arising from 20 terminals (i.e. 0·25 erlang/terminal). With the increased number of terminals there is now a finite probability that more than 10 terminals are concurrently busy. The overall effect is to spread the probability distribution as may be seen in the figure.

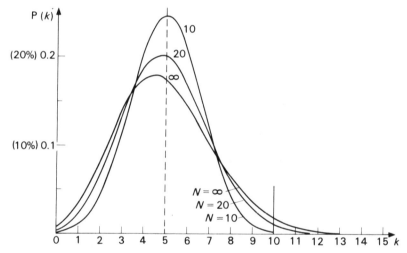

Figure 4.1 Probabilities of k concurrent calls for mean of five (N no. of sources).

In general if A erlangs arise from N terminals the probability distribution is given by

$$P(k) = \binom{N}{k}\left(\frac{A}{N}\right)^k \left(1 - \frac{A}{N}\right)^{N-k}$$

Now as $N \to \infty$

$$\left(1 - \frac{A}{N}\right)^{N-k} \to e^{-A} \quad \text{and} \quad \binom{N}{k} \to N^k/k!$$

so, for large N, the probability distribution of k is

$$P(k) = \frac{A^k}{k!} e^{-A} \qquad (4.3)$$

This is called the Poisson distribution, which is also shown in figure 4.1.

Note that implicit in the derivation of the above distributions is the assumption that calls from the terminals are statistically independent. This is unlikely to be true in a circuit switched system since there will be periods of congestion during which no new calls can be accepted. During these times the probability of a free terminal originating a new call and it being accepted is zero rather than p. In systems where call blocking takes place it is necessary to take into account the effect of blocked calls. This may be illustrated by considering the simplest case of two terminals with access to a single trunk. Using the terminology of queueing theory, the terminals will be called *sources* and the single trunk will be called a *server* [1].

We shall assume that each of the two sources has an average occupancy of 0·4, as shown in Figure 4.2a and 4.2b. The total traffic offered from these sources is therefore 0·8 erlang. According to the binomial distribution, the probabilities of there being 0, 1 or 2 concurrent calls is:

$$P(0) = \binom{2}{0} 0\cdot6^2 = 0\cdot36$$

$$P(1) = \binom{2}{1} 0\cdot4 \times 0\cdot6 = 0\cdot48$$

$$P(2) = \binom{2}{2} 0\cdot4^2 = 0\cdot16$$

So, for 36% of the time (on average) neither source will need serving; for 48% of the time one or other will need serving; and for 16% of the time both will need serving. However, there is only one server available to these two sources, and therefore congestion will occur. There are two ways of specifying congestion:

(a) *Time congestion* — the probability that all servers are busy.
(b) *Call (or demand) congestion* — the proportion of calls arising that do not find a free server.

The difference between time and call congestion exists because, although all servers may be busy, it does not affect a call until a new one arises. If the number of sources is equal to the number of servers, the time congestion is finite, but the call congestion is zero. However, in most practical cases the difference between time and call congestion is not significant, and mathematic-

BASIC EQUATIONS 89

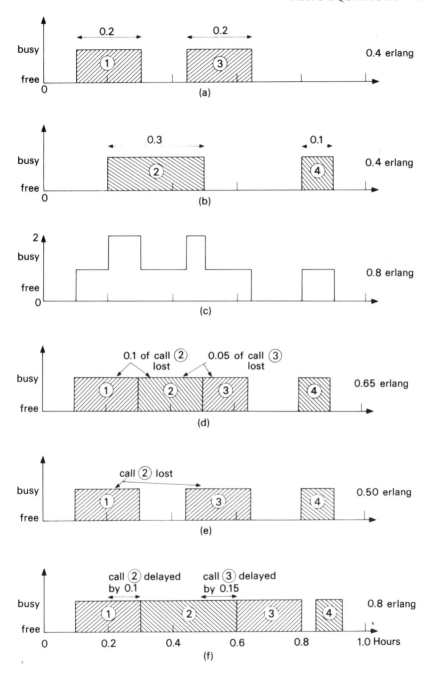

Figure 4.2 Examples of different terminals. (a) Terminal 1. (b) Terminal 2. (c) Number of concurrent calls. (d) Blocked calls held (BCH). (e) Blocked calls lost (BCL). (f) Blocked calls wait (BCW).

90 TRAFFIC THEORY

ally it is easier to compute time congestion. When the number of sources is large, the probability of a new call arising is independent of the number already in progress and therefore the call congestion is equal to the time congestion. The time congestion in our example is:

$$B_T = P(1) + P(2) = 0\cdot 64$$

The computation of the call congestion is more complex and is shown in Appendix B.4.

Where the number of sources is large (so the probability distribution is a Poisson distribution) the time congestion becomes

$$B_T = \sum_{x=M}^{\infty} \frac{A^x}{x!} e^{-A} \qquad (4.4)$$

where A is the total offered traffic.

Equation 4.4 is shown as the Molina equation [2] after the American who first applied it to telephone theory. It is still the basis of much of the practical traffic engineering in North America. Another name for these equations is the *blocked calls held* traffic equations. The reason for this name is the physical interpretation of the mathematical assumption behind the above equation. The implication is that a blocked call continues to demand service for a period equal to the average holding it would have had if successful. If, during this period, a server becomes free, this server is seized and rendered unavailable for the unelapsed part of the holding time (but the call is still regarded as lost). This is of course an unrealistic assumption but, as is seen in the next section, the result is to yield a slightly lower traffic capacity for a given probability of blocking and this is regarded as a useful margin of safety in North America.

Figure 4.2d shows the effect of blocked calls held for our example of the two sources. When the second source's first call starts the server is already occupied. However, the call occupying the server ends before the second call would have ended. This second call uses the server for the duration of its unexpired holding time. Calls 2 and 3 lose $0\cdot 1$ and $0\cdot 05$ units of time respectively and the traffic on the trunk is only $0\cdot 65$ erlang; $0\cdot 15$ erlang of traffic is therefore lost.

Erlang equations. An alternative assumption to that of blocked calls held is that, if a call arises when all servers are busy, the call is immediately cleared (with zero holding time) and does not reappear during the period of observation. This is known as the *blocked calls lost* assumption. The probability distribution giving the number of current calls in this case was first worked out by A. K. Erlang [3]. The implications of this assumption are illustrated in Figure 4.2e. Call 3 from the second source arises when the server is busy, so it is cleared immediately. When call 2 from the first source arises the server is free, so this call proceeds. In this case the total traffic carried by the server is $0\cdot 50$ erlang; $0\cdot 30$ erlang is lost.

BASIC EQUATIONS 91

Appendix B (Equation B.8) shows that, when there are a large number of sources with access to M servers, the probability of there being k servers busy is given by:

$$P(k) = \frac{\dfrac{A^k}{k!}}{\displaystyle\sum_{x=0}^{x=M} \dfrac{A^x}{x!}} \qquad (4.5)$$

where A is the total traffic produced by the sources. This is known as the *Erlang-B* distribution and is perhaps the best known and widely used distribution in traffic theory. This is because in a large number of practical cases, the assumption that there are a large number of sources and the assumption that lost calls are cleared represent a realistic model of the system. The distribution can be shown to be valid for any statistical distribution of the holding times [4].

The probability of blocking $E(M, A)$ where there are M servers and an offered traffic of A erlangs is given by putting $k = M$ in Equation (4.5)

$$E(M, A) = \frac{\dfrac{A^M}{M!}}{\displaystyle\sum_{x=0}^{x=M} \dfrac{A^x}{x!}} \qquad (4.6)$$

This is called the *Erlang formula of the first kind*, and gives the time congestion of a system. This formula relates the probability of a blocked call to the offered traffic and the number of servers. For design purposes it is necessary to know the inverse relationship between the number of servers required for a given offered traffic and the required probability of blocking. There are extensive tabulations of this relationship [5] (see, for example, Fig. 4.3).

Any call arising while the M servers are all busy is assumed to be lost. As the offered traffic is A erlangs, the lost traffic is $AE(M, A)$ erlangs.

Engset equations. For the case where the number of sources is limited to N and blocked calls are lost, the probability of k servers being busy is given by the Engset equation (Equation B.11 of Appendix B):

$$P(k) = \frac{\dbinom{N}{k} \rho_m^k}{\displaystyle\sum_{x=0}^{M} \dbinom{N}{x} \rho_m^x} \qquad (4.7)$$

where ρ_m is the probability that a new call arises from a free terminal, i.e.

$$\rho_m = \frac{A}{N - A} = \frac{\rho}{1 - \rho}$$

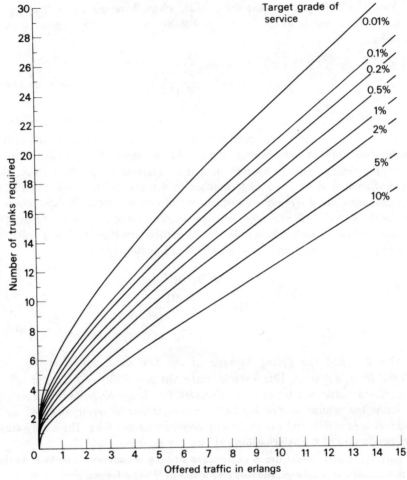

Figure 4.3 Traffic capacity on basis of Erlang-B formula.

It can easily be shown that the Engset equation of Equation 4.7 is the general case of the Erlang-B and the binomial distribution since:

(a) If $N \to \infty$ the Engset equation (4.7) tends to the Erlang-B equation (4.5).
(b) If $N = M$ the Engset equation becomes the binomial distribution (4.2).

4.2 Queueing systems

Blocked calls wait (Erlang-C formula). If the switching system has some mechanism whereby a blocked call can wait until a server becomes free, the probability of a call being delayed (Equation B.19 of Appendix B) is:

COMPARISON OF THE VARIOUS DISTRIBUTIONS

$$B = \frac{\dfrac{A^M}{M!} \dfrac{1}{1-\eta}}{\displaystyle\sum_{x=0}^{M-1} \dfrac{A^x}{x!} + \dfrac{1}{1-\eta} \dfrac{A^M}{M!}} \quad (4.8)$$

where η is the average occupancy of the servers, that is $\eta = A/M$. This is known as the *Erlang-C formula* and is valid only with a queue of infinite length and an infinite number of sources.

It is shown in the references [6] that in such a queueing system if delayed calls are served in a first-in first-out order, the probability, $W(\tau)$, of a call being delayed more than τ seconds is given by

$$W(\tau) = \exp\{-m(1-\eta)\tau\} \quad (4.9)$$

where $m = M/\eta$. This distribution is a negative exponential with a mean value of

$$\overline{W(\tau)} = \frac{1}{m(1-\eta)} \quad (4.10)$$

$$= \frac{h}{M(1-\eta)}$$

i.e. the mean value is the $(1/M)$th of the mean holding time divided by $1 - \eta$ where η is the average occupancy of each server. Note that, as the server occupancy approaches unity, the average waiting time approaches infinity.

This result is obtained for a first-in first-out (FIFO) queue. If different queue philosophies are adopted such as last-in first-out (LIFO) or service in random order (SIRO) the distribution of waiting times is different. However, the mean time will be the same for these other cases.

4.3 Comparison of the various distributions

It is instructive to compute some specific results based on the equations discussed above. Table 4.1 gives some results for the case of two servers with 1 erlang of offered traffic. Table 4.2 considers further the case of blocked calls wait, assuming a negative exponential distribution of call holding times with a mean of 200 seconds.

From the simple examples above it may be seen that the different assumptions about the behaviour of users produce significant differences in both the probability distribution of the number of occupied servers and the lost traffic. The cases with a limited number of sources have lower time congestion and less traffic loss than when the same traffic is offered from a very large number of sources. This is to be expected because, when the number of sources is limited, the probability of a new call arising decreases as the number of busy servers

94 TRAFFIC THEORY

Table 4.1 Some results for a two-server system

Conditions	Time congestion (%)	Lost traffic (erlangs)	Probability (%) of occurrence of a number of calls	
			Zero	One
The traffic is produced by two sources	25·0	0	25·0	50·0
The traffic is produced by four sources and blocked calls are held	26·2	0·16	31·6	42·2
The traffic is produced by a very large number of sources and blocked calls are held	26·4	0·26	36·8	36·8
The traffic is produced by four sources but blocked calls are lost	22·2	0·11	33·3	44·4
The traffic is produced by a very large number of sources and blocked calls are lost	20·0	0·20	40·0	40·0
The traffic is produced by a very large number of sources but blocked calls wait	33·3	0	33·3	33·3

Table 4.2 Results for two-server system with infinite queue (1 erlang from infinite number of sources)

Condition	Parameter
Mean waiting time for blocked calls	200 seconds
Mean waiting time for all calls	67·7 seconds
Probability that a blocked call has to wait less than 200 seconds	36·8%
Probability that any call has to wait less than 200 seconds	12·3%
The maximum time within which 99% of all calls are given service (assuming first come-first served)	682 seconds

increases. With a very large number of sources the probability of a new call arising is independent of the number of calls already in progress.

Comparing the time congestion figures for the three assumptions of user behaviour shows that the time congestion is highest for blocked calls wait and lowest for blocked calls lost. However, the lost traffic is greatest for blocked calls lost.

The explanation for this is simple. If a call arises when all servers are busy, under the blocked calls lost assumption that call is lost and does not reappear. When one of the servers becomes free it remains free until a new call arises.

However, under the blocked calls wait assumption, as soon as a server becomes free it is used by a waiting call if blocking has occurred. This means that the proportion of time the system has all servers busy is higher but more traffic is carried.

The blocked calls held assumption is midway between these two assumptions; it represents a form of queueing because the mathematical implication of the assumption is that a blocked call 'waits' until its natural holding time is completed and, if during this time a server becomes free, it will use that server.

The difference between the results of the three assumptions increases as the average load per server is increased. Figure 4.4 shows the blocking probability computed under the three assumptions for a range of different loads.

The choice of distribution for design work depends upon the system characteristics. If the system (or the user) has the capability for waiting, Erlang-C may be used. The blocking probabilities are valid for any distributions of holding time, but the distribution of waiting times depends on the actual holding time distribution and also on the queue service rule (first come/first served, random order, or last come/first served). Formulae have been evaluated for most practical alternatives. A commonly used simplification are the Crommelin curves [7] for constant holding time.

The choice between blocked calls held or blocked calls lost is more difficult as neither accurately represents typical user behaviour, except in special circumstances (see Ch. 5). In the case of a telephone system, a user often makes frequent repeat attempts to set up a call if the attempts meet congestion. It has been found difficult to characterise mathematically these second attempts [8]. It is thought that the blocked calls held assumption is reasonably valid because, for the same value of time congestion, it gives lower lost traffic than the blocked calls lost assumption. In any case, the differences between the results of calculations based on those different assumptions are outweighed in practice by the errors inherent in estimating the offered traffic [9].

A further complicating factor is that the mean value of the traffic normally rises with time and equipment is provided in blocks to satisfy the new demand. For instance, a particular trunk route may have its capacity extended by blocks of 12 circuits at a time. If a trunk route is expanded annually, the additional equipment will be chosen so that it still satisfies the grade of service objectives just before the next extension occurs. Until that date the grade of service will be better than the design objectives (Fig. 4.5). Thus in a network which is undergoing continual expansion (at different times) the average grade of service is much better than the design objectives [10].

In North America design work has traditionally been based on the blocked calls held assumption whereas European administrations generally favour the blocked calls lost assumption.

The next section shows how these distributions are used in the design of practical switch networks. Chapter 5 shows how they are used in the design of complete telephone networks.

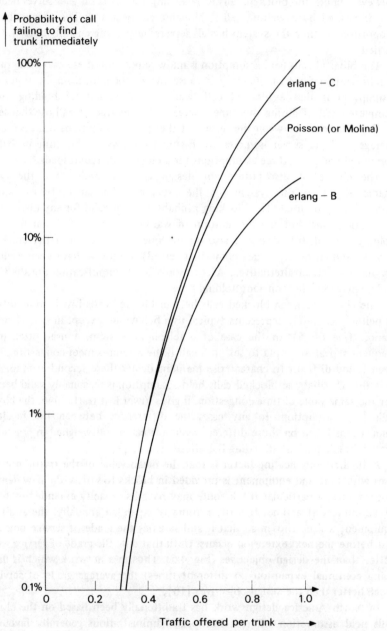

Figure 4.4 Comparison of different traffic theories (for 10 servers).

APPLICATIONS OF TRAFFIC THEORY 97

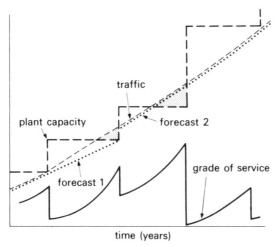

Figure 4.5 Effect of provisioning period on grade of service.

4.4 Applications of traffic theory

The practical application of traffic theory to the dimensioning of resource sharing systems will now be described with reference to Figure 4.6 which shows a number of sources sharing access to a small number, M, of resources called *servers*. The servers might be junctors, registers or trunks to other switching systems. For the moment we shall assume that the switch network can connect any free inlet to any free outlet. This is called a *full availability* switch network.

The method of design is to decide on a target grade of service and then to find the number of servers required to carry the expected level of traffic. For infinite sources and blocked calls lost behaviour, this can be achieved using the Erlang-B formula. Many tables have been produced giving the required number of servers as a function of the offered traffic and the target grade of service. The form of these relationships is shown graphically in Figure 4.4.

It may be seen from Figure 4.4 that all the curves tend to straight lines once the traffic offered exceeds a few erlangs. Useful approximations for the traffic capacities are:

$$T = 5 \cdot 5 + 1 \cdot 17 \, E \text{ for } B = 1\%$$
$$T = 7 \cdot 8 + 1 \cdot 28 \, E \text{ for } B = 0 \cdot 1\%$$
(4.11)

where T is the number of servers and E is the offered traffic in erlangs. The approximations are accurate to ±1 server for $5 < E < 50$ erlangs.

Traffic carried per server. It is useful to look at some specific examples of the above functions. For an offered traffic level of 5 erlangs, 7 servers are needed to give a 10% grade of service. For 0·2% grade of service 13 servers are needed.

98 TRAFFIC THEORY

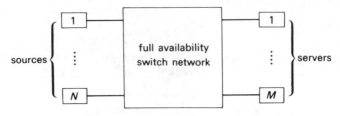

Figure 4.6 Basic model for resource sharing.

Note that to achieve a low probability of blocking the number of servers must be more than double the average number of calls.

The corresponding numbers of servers for 10 erlangs of offered traffic are 13 for a 10% grade of service and 20 for a 0·2% grade of service. So, doubling the traffic intensity does not necessitate doubling the number of servers (for a given grade of service). This is, in fact, a general principle: the higher the intensity of offered traffic, the more traffic is carried by each server. For a 0·2% grade of service, 13 servers carry 5 erlangs (an average loading of 0·38 erlang per server) whereas 20 servers can carry 10 erlangs (an average loading of 0·50 erlang per server).

Figure 4.7 is a graph of average traffic carried per server against the total traffic offered to the servers; the graph is applicable when the number of servers is chosen to give a grade of service of no more than 1% in all cases. Note that the traffic carried per server tends to unity as the offered traffic increases (Table 4.3), although the approach to unity is very gradual; even for 200 servers, the average traffic is still only 0·90 erlang per server.

The above discussion does not take into account the effect of traffic overload. In designing a switching network or other part of a switched system it is necessary to ensure that it will still operate satisfactorily under a moderate overload. Figure 4.8 shows the effect on the blocking probability of various overloads for different sized groups of servers. For each size group the original traffic is that which may be carried at a 1% grade of service. It can be seen that a 25% overload on a 5-server group leads to a worsening of the grade of service from 1·0% to 2·1%. The same overload on a 50-server group causes the grade of service to worsen to 7·7%. The larger, more efficient group is therefore more sensitive to overload. Above a particular traffic level the number of servers needed is determined by the required overload performance rather than the basic traffic. For instance, a typical requirement might be that a normal grade of service of 1·0% should increase to no more than 5% for a 10% overload. So there would be no advantage in having groups of more than about 35 servers for this required grade of service. This corresponds to an average loading of only 0:70 erlang per server. If the server group were larger than 35 the number provided would be governed by the overload requirement and the average loading could not be increased.

APPLICATIONS OF TRAFFIC THEORY 99

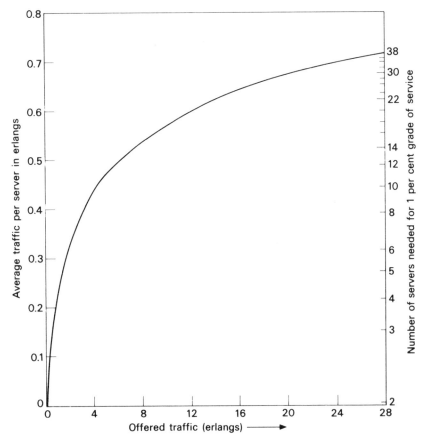

Figure 4.7 Average traffic carried per trunk for 1 per cent grade of service.

Table 4.3 Relationship between offered traffic level and server efficiency

No. of servers	Capacity in erlangs for 1% grade of service	Average traffic carried (erlangs/server)
10	4·5	0·45
20	12·0	0·60
50	37·9	0·76
100	84·1	0·84
200	179·7	0·90

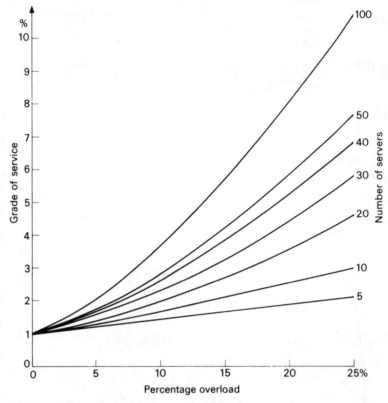

Figure 4.8 Effect of overload on grade of service.

To give another example, if the requirement were for a normal grade of service of 0·2% and a worsening of no more than 1·0% under 10% overload, a 70-server group would give the maximum allowable efficiency. In this case the maximum loading would be 0·73 erlang per server.

In general it is not possible to load servers to an average of more than about 0·70 to 0·75 erlang (70–75% efficiency) without jeopardising the overload performance. (Note that traffic is not necessarily distributed evenly between servers, so particular servers may carry close to 1 erlang while others carry significantly less.)

Use of restricted availability switches. It was shown above that a large group of servers' trunks is needed to obtain a server utilisation approaching 0·8 erlang per server. This implies the use of large numbers of crosspoints in a switching network. For example, consider a group of 200 sources generating a total of 10 erlangs of traffic. Assume that a grade of service of 0·2% is required. To carry the 10 erlangs, 20 servers are needed. A matrix of 200 × 20 crosspoints is therefore needed, that is 20 crosspoints per source (see Fig. 4.9a).

APPLICATIONS OF TRAFFIC THEORY 101

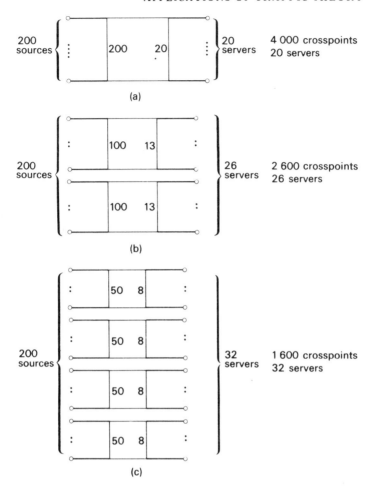

Figure 4.9 Use of restricted availability switches to reduce total number of crosspoints. (a) One group. (b) Two groups. (c) Four groups.

With many electro-mechanical switching devices, such as reed relays or crossbar switches, this size of switch is uneconomic. The number of crosspoints can be reduced if the sources are split into two groups, each of which generates 5 erlangs (Fig. 4.9b). Each group now requires 13 servers, so the total number of crosspoints is reduced to 2600. However, 26 rather than 20 servers are now needed. If the sources are split further, into four groups, each group requires 8 servers. In this case there are 1600 crosspoints and 32 servers (Fig. 4.9c).

So, switches with smaller numbers of outlets may be used if the traffic is split into smaller amounts. However, the non-linear nature of the traffic curves implies that small amounts of traffic are carried much less efficiently than larger amounts. Switch networks which do not allow an inlet access to all possible servers are referred to as *restricted availability* switch networks (as distinct from *full availability* networks). The term *limited availability* is sometimes used, but its meaning gives rise to confusion when applied to multi-stage networks.

Although it is generally inefficient in the use of servers to split the sources into completely separate groups (in order to reduce the number of crosspoints), it is possible to achieve reasonably high efficiency by employing a certain approach. In order to understand what this involves, consider a switch whose control causes it to search sequentially for a free server. (An example of this is a rotary switch, such as a uniselector or the horizontal bank of a two-motion selector.) When the control is requested to find a free server it always starts searching in the same order and starts from the same server. If the total traffic offered to the servers is A erlangs, all this traffic is offered to the first server. The traffic offered to the second server is the traffic lost to the first server, that is:

$$AE(1, A) = A \frac{A}{1+A} = \frac{A^2}{1+A}$$

where $E(1, A)$ is the Erlang-B blocking probability for $M = 1$ (Equation 4.6). The traffic carried by the first server is the difference between the offered traffic and lost traffic:

$$A[1 - E(1, A)] = \frac{A}{1+A}$$

This computation may be continued with the other servers. In general it can be shown that the traffic carried by the nth server in a sequentially hunting, blocked call lost system is given by:

$$A_n = A[E(n-1, A) - E(n, A)] \qquad (4.12)$$

For instance, if a traffic level of 5 erlangs is offered sequentially to a number of servers, the traffic levels carried by the first three servers are 0·833, 0·788 and 0·730 erlang.

Figure 4.10 shows some typical distributions of traffic carried by servers with sequential offering of calls to the servers. The traffic carried by the later choice servers is small, but their presence is essential to provide the required grade of service. It should be noted that, if random hunting is used, the same traffic is carried by each server, but the total traffic carried by a group is unchanged.

If, for reasons of economy, there is restricted availability to the servers, there are several groups of trunks all with late choice servers carrying low traffic levels. It is an obvious step to combine these late choice servers as shown, for instance, in Figure 4.11 where the last two servers on each of the two groups are shared. Exact calculations of the grade of service for such networks is difficult, but several techniques exist to estimate it and are discussed in the references [11].

APPLICATIONS OF TRAFFIC THEORY   103

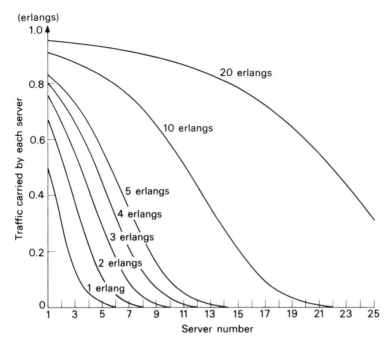

Figure 4.10   Traffic distribution for sequential hunting.

As a specific result it may be estimated that the network of Figure 4.11a can carry 11·5 erlangs at a 0·2% grade of service.

If the number of outlets available to each group is reduced to 12, the traffic capacity is about 10·1 erlangs and only 22 servers are needed (Fig. 4.11b). In order to obtain accurate performance figures it is necessary to use switch network simulation as will be outlined later.

This process of combining the later choice trunks is called *grading* and is a common feature of many types of switching network. In practice, grading may be extended to several groups of sources. For instance, Figure 4.11c shows a four-group arrangement where each group has access to ten servers. The fourth to seventh choices are commoned in pairs and the eighth to tenth are commoned in quadruples. This example has a total of 23 trunks. It may be estimated that this network can carry 9·9 erlangs at a 0·2% grade of service. Had the trunks been provided on a fully available system, 20 servers would have been required for the same grade of service.

Grading can be used in systems other than those with sequential hunting. The same traffic will be carried by the network of Figure 4.11c if the search for a free server is made in groups, that is if the first choice group (1—3) is tested first followed by the second choice group (4—7) and finally the last choice (8—10). Grading patterns are known which allow for completely random searching [12]

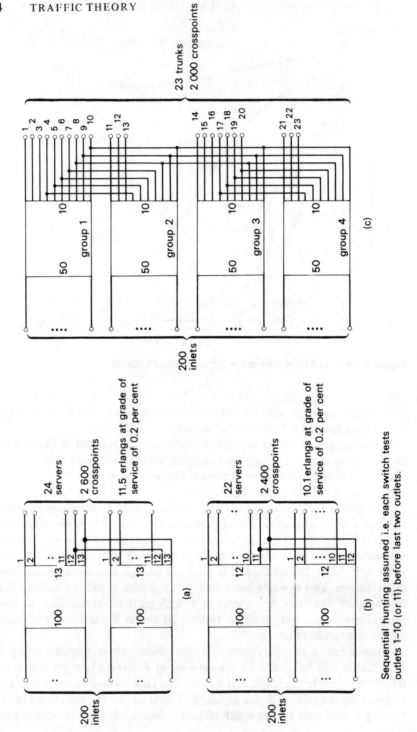

Figure 4.11 Simple examples of grading. (a) 13 outlet groups. (b) 12 outlet groups. (c) Four-group grading.

MULTI-STAGE SWITCH NETWORKS – CONDITIONAL SELECTION 105

but these are often inconvenient in practice as the wiring between switches becomes complex.

4.5 Multi-stage switch networks – conditional selection

So far we have discussed the use of restricted availability switches in a single switching stage to provide access from a number of sources to a number of servers.

An alternative way of using restricted availability switches is to arrange the switches in two or more stages. For instance, Figure 4.12 shows three different ways in which 200 inlets may be connected to 24 outlets. This arrangement has the property that there is effectively full availability: each inlet has access to each outlet. However, there is now the possibility of internal blocking where an inlet cannot be connected to a free outlet if the only links which give it access are already occupied by other calls. The probability of a call being lost due to *internal* blocking is sometimes called the *mismatch loss*.

In order to get the best out of this type of network it is necessary to select an inter-stage link which has access to a free outlet in the second stage. This is referred to as *conditional selection*. If a free link is chosen at random without regard to whether it has access to a free outlet, the probability of blocking is greater than it would be with conditional selection.

If $B(M-x)$ is the probability that there are exactly x free outlets and $P(x)$ is the probability that there is no free path from a specific inlet to any of the x free outlets, the overall blocking probability is given by:

$$B = B(M) + P(1)B(M-1) + P(2)B(M-2) + \ldots + P(M)B(0)$$
$$= \sum_{x=0}^{x=M} B(x)P(M-x)$$
(4.13)

The coefficients $B(M-x)$ are dependent upon only the traffic carried and not upon the structure of the network. If the number of traffic sources contributing to the offered traffic is large, these coefficients are given by the Erlang-B distribution. In particular, $B(M)$ is the value of the full availability case and the additional terms represent internal blocking.

In simple cases, analytic solutions may be found to Equation (4.13) with the aid of combinational algebra [13]. These are outlined in Appendix E which deals with the more general case shown in Figure 4.13. This case consists of a switch network which provides access from a number of sources to a number of different groups of servers. (In a telephone switch network, this case is the group selector network which provides access from its inlets to a number of different routes, each route having several circuits.)

Figure 4.13 shows a network consisting of K first-stage switches each providing access from n sources to L second-stage switches. The second-stage switches

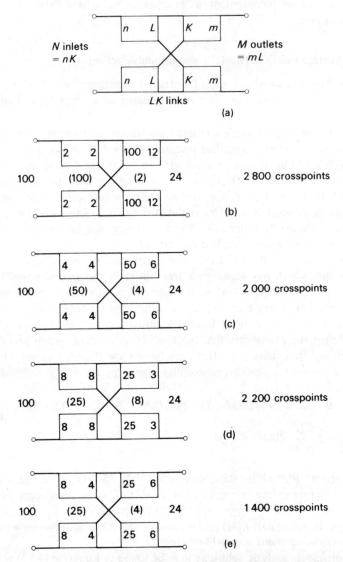

Figure 4.12 Two-stage group selectors. (a) General case. (b) – (d) No concentration in first stage. (e) 2:1 concentration.

provide access to Q groups of servers, each second-stage switch accesses m_1 servers in the first group, m_2 in the second group and so on. The relevant grade of service for this network is the group blocking probability. This is the percentage of time that an inlet has no access to any free circuit in a particular group. Appendix E shows that good approximations to this probability are

MULTI-STAGE SWITCH NETWORKS – CONDITIONAL SELECTION   107

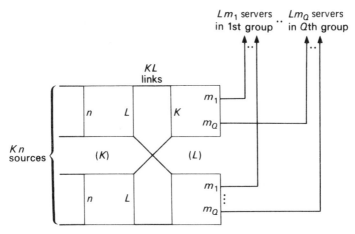

Figure 4.13   General case of group selector.

(a) where there is no concentration in first-stage switch (that is $n = L$)

$$B = \frac{E(M, A)}{E\left(M, \dfrac{A}{\rho}\right)} \qquad (4.14)$$

(b) where there is concentration in first-stage switch (that is $N > L$)

$$B = \rho^L + \frac{E(M, A)}{E\left(M, \dfrac{A}{\rho}\right)} \qquad (4.15)$$

where M is the total number of outlets in the group of outlets being considered and A is the total offered traffic to that group. The average traffic per terminal to *all* outlets (not just the group being considered) is ρ.

If conditional selection is not assumed the blocking probability is roughly the sum of the following two terms:

(a) the probability of finding no free links,
(b) the probability that all of the set of m particular outlets accessible by a chosen link are busy, that is

$$B = P(L) + \frac{E(M, A)}{E(M - m, A)} \qquad (4.16)$$

108   TRAFFIC THEORY

Table 4.4   Blocking probabilities of networks shown in Figure 4.12

Figure	A-stage size	B-stage size	No. of crosspoints	B (%) conditional selection	B (%) un-conditional selection
4·12 (b)	2 × 2	100 × 12	1800	1·66	3·3
(c)	4 × 4	50 × 6	2000	1·02	9·8
(d)	8 × 8	25 × 3	2200	0·98	26·2
(e)	8 × 4	25 × 6	1400	1·53	9·8
	Full availability		4800	0·84	

Table 4.4 shows the group blocking probability for the examples of Figure 4.12b–e in which there are 15 erlangs of offered traffic (i.e. $\rho = 0.075$).

It should be noted that although the use of a two-stage group selector can be an efficient way of using switches, it requires more complicated control mechanisms because of the need to provide conditional selection.

When the number of sources is greater, or when the traffic per source is greater, even more economy in crosspoints and servers may be achieved by the use of three or more stages of switching. This point is discussed in Chapter 7.

4.6   Simulation techniques

In the majority of practical switch network design, it is necessary to use simulation in order to evaluate the traffic carrying capacity because the analytic methods are too complex. Simulation can take into account normal conditions and a range of abnormal conditions such as overload traffic inbalance (more traffic appearing on one set of inlets than on others), equipment faults. Until the early 1960s these simulations had to be performed manually or with specially built equipment, but today they are performed by general purpose computers.

There are three classes of technique used for computer evaluation of traffic capacity:

(a) Direct simulation using *time-true* and *roulette wheel* techniques.
(b) Estimation from channel graphs (discussed in Ch. 7).
(c) Evaluation of mathematical series or solution of simultaneous equations obtained by theoretical analysis.

Time-true simulations.   A time-true simulation technique is the straightforward one in which the network being simulated is modelled as a digital pattern inside a computer memory [14]. For example, there may be a computer word for every row of a switch network and a bit set to a 'one' or a 'zero' to indicate whether a particular crosspoint is in use or not. With the aid of a random number generator, a list is produced of the times at which calls start and finish. It is possible to make the inter-arrival times and holding times follow any desired probability distribution. The lists of these events are placed into time order and the simulation is run by using a program to read the lists. Each time the program finds a call arrival, it attempts to find a path through the network modelled in the memory. When the program finds a call-finished event it clears the appropriate bits in the memory.

It is possible to simulate a very large number of events, to count the number of calls made and to count the number failing because of congestion. It is also possible to produce histograms of waiting times.

This technique requires a very large number of simulated calls in order to achieve a steady state, but yields results with a high degree of reliability. It also requires a large memory in the computer since comprehensive records must be kept of past and future events. For instance, there must be an explicit record of which path each successful call has used.

Roulette simulation. In situations where the call arrivals can be described by a Poisson distribution and call holding times by negative exponential distributions some considerable simplification may be made in the simulation. This simplified method, which is attributed to Kosten [15], relies on the fact that a negative exponential distribution implies that the probability of a call ending at any instant is independent of the previous duration of that call. This means that the call ending event is a pure random event.

The principle of the roulette simulation may be seen from a simple example. Consider the grading system shown in Figure 4.14a which shows two groups of sources, each with access to two servers. Each group has its own server and a third server is the common second choice of each group. Assume that the probabilities (per unit time) of call arrival in the two groups are λ_1 and λ_2 and that the mean holding time of calls is h. For a negative exponential distribution of holding times, the probability of a call ending is $1/h$ per unit time (independent of the length of time the call has already been in progress).

The setting up and clearing down of calls may be described by the sum of five random processes: two arrival processes with mean arrival rates λ_1 and λ_2 and three departure events each with a mean rate of $1/h$. These may be combined into a single random process with mean of $\lambda_1 + \lambda_2 + 3/h$ events per unit time and the proportion of the events that occur will be in the ratio $\lambda_1 : \lambda_2 : 1/h : 1/h : 1/h$. A roulette wheel may be imagined as in Figure 4.14b, having its circumference split into five areas in the ratios $\lambda_1 : \lambda_2 : 1/h : 1/h : 1/h$. As the wheel is spun, a series may be produced which gives the arrival and departure events in the correct proportions. These events will be in a representative sequence and so this technique is often referred to as a *sequence-true* simulation, that is there is no mention of the time between the events.

In order to run a simulation in a computer, random numbers are generated to access a table with the required probability distribution. This indicates whether

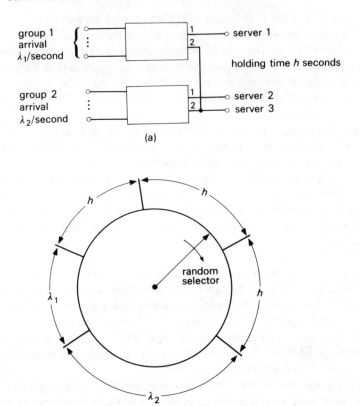

Figure 4.14 Principle of roulette simulation. (a) Simple grading system. (b) Roulette wheel. Selector rotates and stops at random and arc. When it stops it indicates next event.

the next event is an arrival or a departure. If the event is an arrival the appropriate server is marked as busy. That is if the arrival is on group 1, server 1 is marked as busy if it is previously free, otherwise server 3 is busied. If both are busy, the event is counted as a lost call. A departure event from a particular server ensures that the server is marked as free. If it is already free, the event is ignored.

It may be seen from this description that the great advantage of the roulette method is that there is no need to construct lists of events. All that has to be stored is the state of the network. It is possible to obtain waiting time distributions by this method, but not estimates where time appears explicitly, for example the probability of a request having to wait longer than a certain time.

4.7 Traffic measurement and traffic prediction

Finally it must be pointed out that the theories discussed in this chapter assume that the value of offered traffic is known accurately. In practice there are errors in the measurement of traffic and in the extrapolation of traffic figures to estimate the future load upon switching centres. The traffic measurement errors arise first because the traffic itself is a random quantity and second because it is measured by a sampling procedure.

Typically, traffic measurement is performed by regularly counting the number of servers in use over the busy hour. For instance, the number of registers in use would be sampled every 10 seconds while the number of trunk circuits on a particular route would be counted every 3 minutes.

These errors and their effect on the planning process are discussed extensively in the literature [16].

References

1. General queueing theory is well covered in numerous texts. They include:
   Kosten, L. (1973) *Stochastic Theory of Service Systems*. Pergamon Press.
   Kleinrock, L. (1974) *Queueing Systems, Vol. I*. Wiley.
   Beckmann, P. (1968) *Introduction to Elementary Queueing Theory and Telephone Traffic*. Golen Press, Boulder, Colo.
2. Molina, E. C. (1927) The theory of probabilities applied to telephone trunking problems, *Bell Syst. Tech. J.*, 6, pp. 461–94.
3. Brockmeyer, E. *et al.* (editors) (1948) The life and works of A. K. Erlang, *Trans. Danish Acad. Tech. Sci. (in English)*, 2.
4. For example, Kosten, L. (op. cit.) Chapter 4.
5. For example, Brockmeyer, E. (op. cit.).
   Also *Telephone Traffic Theory Tables and Charts* – Part 1. Published by Siemens Aktiengesellschaft with introduction in German and English. This book also contains tables of the Engset formula and tables for use in dimensioning overflow routes (discussed in Chapter 5 of this book).
6. For example, Kosten, L. (op. cit.) Chapter 2.
7. Crommelin, C. D. (1933) Delay probability formulae, *Post Office Elect. Eng. J.*, 26, pp. 266–74.
   Riordan, J. (1953) Delay curves for calls served at random, *Bell Syst. Tech. J.*, 32.
   Design charts and tables may be found in reference 5.
8. Closs, C. and Wilkinson, R. I. (1952) Dialling habits of telephone subscribers, *Bell. Syst. Tech. J.*, 31, pp. 32–67.
9. Wilkinson, R. I. (1971) Some comparisons of load and loss data with current teletraffic theory, *Bell. Syst. Tech. J.*, 50, pp. 2808–32.
10. Povey, J. A. (1975) Teletraffic engineering, in *Telecommunications Networks*. J. Flood, ed., Peter Peregrinus.
11. For example, see a review in Bear, D. (1976) *Principles of Telecommunication Traffic Engineering*. Peter Peregrinus.
12. Elldin, A. (1957) Further studies on gradings with random hunting, *Ericsson Technics*, 55, pp. 177–257.
13. Jacobaeus, C. (1950) A study on congestion in link systems, *Ericsson Technics*, 48.
    Elldin, A. (1967) *Automatic Telephone Exchanges with Crossbar Switches – Switch Calculations*. L. M. Ericsson.
14. For example, Beastall, H. and Povey, J. A. (1973) Teletraffic studies of TXE 4, *Post Office Elect. Eng. J.*, 65, p. 251.

15. This method is described in Chapter 9 of Kosten, L. (1973) *Stochastic Theory of Service Systems*. Pergamon.
16. For example, Chapters 12 and 13 of Bear, D. (1975) *Principles of Telecommunication Traffic Engineering*. Peter Peregrinus.

    Descloux, A. (1965) On the accuracy of loss estimates, *Bell Syst. Tech. J.*, **44**, pp. 1139–64.

    Hayward, W. S. (1952) On the reliability of telephone traffic load measurements by switch counts, *Bell Syst. Tech. J.*, **31** pp. 357–77.

    Iverson, V. B. (1973) Analysis of real teletraffic processes based on computerised measurements, *Ericsson Technics*, **29**, pp. 1–64.

    Iverson, V. B. (1976) On the accuracy in measurements of time intervals and traffic intensities with application to teletraffic and simulation, Report 26, Institute of Mathematical Statistics and Operations Research, Technical University of Denmark.

5
Telephone network organisation

5.1 Network planning

The initial building, and later growth, of a telecommunication system comprising a network of switching centres and transmission links, involves the drawing up of plans for

>routing,
>numbering,
>signalling,
>transmission,
>charging,
>network control, and network administration.

These plans are interdependent and are affected by the predicted (or planned) growth rate of the telecommunication system. The choice of a plan for a telecommunication system generally involves comparison of the economics of various possible plans, but it may also involve a certain amount of human judgement where the various relevant factors cannot be reduced to purely economic terms. The first part of this chapter describes and compares the basic routing plans that are possible and, in particular, the general planning of alternative routing strategies. Also, local, national and international numbering plans are discussed.

To provide flexibility, so that the routing of a call may be chosen independently of the numbering scheme which defines the call destination, it is necessary to provide intelligence at some, if not all, of the switching centres. A later section of the chapter explains how this is achieved and, in effect, defines the most general requirement for a switching centre. However, not all switching centres can economically provide the full translation facilities required for full flexibility. This is especially true of directly controlled step-by-step systems. Where these are used in a network, some compromises have to be made in the flexibility requirement.

114 TELEPHONE NETWORK ORGANISATION

Section 5.6 discusses the philosophy and techniques for billing users for telephone calls. Finally, Section 5.7 discusses the requirements and techniques for overall network management.

5.2 Routing plans

Mesh network. When there are only a small number of switching centres, a simple routing arrangement is the provision of direct routes interconnecting all centres. Figure 5.1a shows such a mesh of trunk routes for four centres, where there is a total of six routes. As a simple example, assume that each centre generates 1 erlang of traffic to each of the other centres. The number of trunks required may be found from Figure 4.3. This shows that four trunks in each direction can carry 1 erlang at just over 1% grade of service. However, if switching centres are able to use the trunk circuits for either direction of traffic, a total of six trunks is sufficient for a grade of service of just over 1%.

For low traffic levels (below about 10 erlangs) the use of two-way, rather than one-way working provides a significant reduction in the number of trunk circuits required. However, two-way working increases the cost of a switching centre. On high traffic routes a compromise is often used where the circuits are divided into three groups. The trunks between two centres are split into two groups of one-way trunks and a third group of both-way trunks. A centre attempts to find a free trunk from its outgoing group before trying one in the common group.

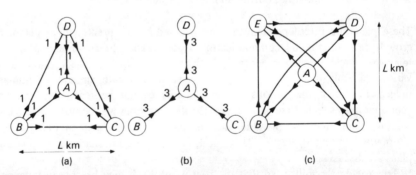

Figure 5.1 Comparison of simple mesh and star connection patterns. (a) Mesh. (b) Star. (c) Five centres mesh (numbers denote traffic flow measured in erlangs).

Star network. An alternative to the mesh arrangement is that of a star, whereby one centre acts as a trunk centre to all the others. Figure 5.1b shows this arrangement for four centres. In this arrangement each route carries a total

ROUTING PLANS

of 6 erlangs; 12 trunks are required on each route for a 1% grade of service. Since most connections require two links connected end-to-end it is necessary to provide a better grade of service for each link. To achieve a grade of service of 0·5% requires 13 trunks. The total number of trunks in this latter case is 39. This is higher than the 36 required for both-way direct routes, but the average route length is less. For instance, if the three outer centres are situated at the corners of an equilateral triangle with a side of length L km, and the fourth centre is equidistant from the other three, the product of distance times trunks is 28·1 L for the mesh arrangement and 20·1 L for the star arrangement. In other words, the distance—trunk product is reduced by 28% with the star arrangement.

Table 5.1 shows the result of similar calculations for different traffic levels and also for five centres. It may be seen that, as the traffic increases, a star connection requires an increasingly larger total number of trunk circuits, but always a lower distance—trunk product. The percentage reduction in distance—trunk product becomes less as the traffic increases and as the number of centres increases.

Some part of the cost of a trunk circuit is a function of its length, but there is also a fixed cost for the trunk control unit at either end plus a suitable share of the switching centre cost. In a local telephone system (based on analogue technology) covering a town, it is the fixed cost which usually predominates. This indicates that the total number of circuits is the significant factor and suggests that mesh is preferable to star connection. However, on long distance networks the fixed cost is less significant and greater economy is obtained by bringing together circuits into fewer routes. In these cases star networks are usually preferable.

The impact on the switching centre design must also be considered. A mesh connection requires more trunk terminations on every switching centre. With the star arrangement the outer centres require fewer trunk terminations, but the trunk centre has a greatly increased number of terminations. In addition, in the star connection, more through traffic has to be switched. So, the star arrangement reduces the design requirements on all but one of the switching centres, but introduces the need for a larger, more powerful trunk centre. Since only one larger centre is required, the net effect on switching costs is often to make the star arrangement preferable.

Mixed routing. In practice a network is hardly ever pure star or pure mesh. A more common arrangement, typical of medium-sized towns, is shown in Figure 5.2a. Here there are a number of direct routes if the traffic justifies it, otherwise the connections are via a trunk centre.

In very large towns, such as London or New York, there are several trunk centres. Such local trunk centres, serving many local centres in a metropolitan area, are referred to as tandem centres (or exchanges). There may be several suburban tandem centres and one large central tandem serving all the centres in a town like London (see Fig. 5.2b).

Table 5.1 Comparison of mesh and star connections (computed for about 1% grade of service for each trunk route)

Traffic from one centre to another (erlangs)	4 centres						5 centres					
	Total trunks		Distance × trunks		Saving on distance × trunks		Total trunks		Distance × trunks		Saving	
	Mesh	Star	Mesh	Star			Mesh	Star	Mesh	Star		
1	36	36	28·1L	20·1L	28%		60	60	57·9L	42·4L	27%	
2	54	60	42·1L	33·5L	20%		90	100	86·9L	70·7L	19%	
3	72	81	56·1L	45·3L	19%		120	136	115·9L	96·2L	17%	

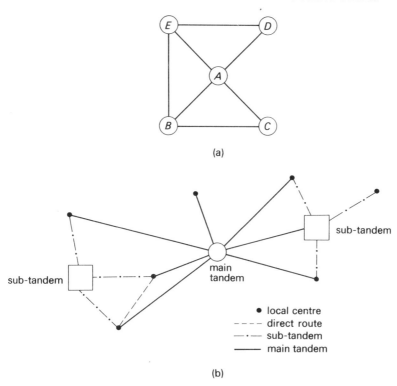

Figure 5.2 (a) Mixed routing. (b) Use of tandem and sub-tandems in large towns.

To calculate whether a direct route is cheaper than a tandem route it is necessary to define a cost ratio, λ:

$$\lambda = \frac{\text{cost of provision of a tandem connection between two centres}}{\text{cost of provision of a direct circuit between two centres}}$$

The costs are usually measured in terms of the *present value of annual charges*. This is an accounting technique whereby all future running costs of a system are discounted to a single sum and added to the capital value of an item of equipment [1].

The cost for a tandem connection (as in Figure 5.2a) includes the cost of trunk circuits *BA* plus trunk circuits *AC*, together with a suitable share of the cost of extra switching required at exchange *A*. Thus, it would appear that λ is always greater than 2, as two trunk circuits are involved in addition to a switching cost at *A*. However, it is often the case that λ is less than 2 because economies of scale mean that the cost of providing an extra trunk circuit to a main tandem may be less than that of providing a geographically direct circuit.

118   TELEPHONE NETWORK ORGANISATION

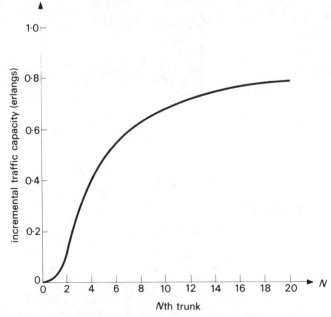

Figure 5.3 Incremental efficiency of adding Nth trunk to group of $N-1$ (1 per cent grade of service).

(For instance the cost per kilometre of a 50-pair cable is typically only three times that of a 10-pair cable.) A further factor which reduces the cost ratio below 2 is that a 'direct' route is often provided by pairs in the main cable connected to the tandem but the pairs are connected across at the tandem. Therefore, an electrically direct circuit may in fact be geographically indirect. In this case the distance cost is the same for both direct and tandem; the difference in cost comes from the two extra trunk control units and the extra switching equipment. In practice cost ratios are usually found in the range 1·2 to 2·5.

Clearly, routing via a tandem switching centre is always more economic if $\lambda \leqslant 1$, but the non-linear relationship between number of trunks and traffic carried can make tandem rather than direct routing more economic even for values of λ greater than unity. This will be shown by an example. At a 1% grade of service a single trunk can carry 0·01 erlang, two can carry 0·15 erlang, and three 0·45 erlang. Thus adding one trunk to a route of only one trunk increases the traffic capacity by only 0·14 erlang. Adding a third trunk to a two-trunk route increases the traffic capacity by only 0·30 erlang. However, the addition of an eleventh trunk to a ten-trunk route increases the capacity by 0·7 erlang. Figure 5.3 shows the incremental traffic capacity produced by the addition of an Nth trunk to a route of $N-1$ trunks for the fixed grade of service of 1%.

It can be seen that, as a general rule increasing the capacity of an existing trunk route always requires fewer additional trunks than the provision of a new, direct trunk route. For example, consider the traffic levels shown in Figure 5.4a.

ROUTING PLANS 119

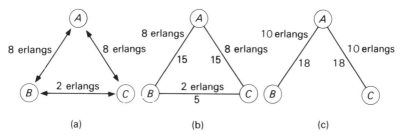

Figure 5.4 Comparison of tandem and direct routing. (a) Traffic flows for example. (b) Direct routing. (c) Tandem routing.

The 8 erlangs on the routes *AB* and *AC* require 15 trunks on each route. The 2 erlangs of traffic on route *BC* require 6 trunks using a direct route (Fig. 5.4b). However, on the main route an additional 2 erlangs can be carried by the addition of only 3 trunks (Fig. 5.4c). So, the tandem route is more economic for a cost ratio, $\lambda \leqslant 2 \cdot 0$.

As a second example, consider a higher traffic level on route *BC* of 10 erlangs. In this case the choice is between 18 direct trunks or 12 extra connections via the tandem switching centre. Tandem routing in this case is cheaper only if $\lambda \leqslant 1 \cdot 5$.

Alternative routing. The discussion above shows that, for particular ranges of cost ratios, it is more economic to route a stream (or *parcel* as it is often called) of traffic via a tandem centre rather than to route it directly between local centres. The discussion was based on the choice of all-direct or all-tandem routing. In fact, even greater economies are often possible if just a proportion of the traffic is routed directly. This approach is known as *alternative routing* (or *alternate* routing in N. America). In this approach a few direct trunks between two centres are provided. These trunks are said to form a *high usage* route. If all these trunks are busy a call between the two centres is routed via a tandem centre (known as the overflow route).

If the high usage route consists of M trunks and the offered traffic is A erlangs, the probability of all trunks busy is given by the Erlang-B formula (Equation 4.6) to be $E(M,A)$. Hence

traffic carried on high usage route

$$A_C = A(1 - E(M,A)) \text{ erlangs} \tag{5.1}$$

overflow traffic

$$A_O = AE(M,A) \text{ erlangs} \tag{5.2}$$

The Erlang-B formula is a good representation of the traffic on a high usage route because blocked calls are diverted to the next choice route and do not therefore reappear as second attempts. Therefore, the basic assumption of blocked calls lost is satisfied. The graphs in Figure 4.10 may be used to obtain numerical values of Equation 5.1.

Let us look again at the example in Figure 5.4a. A_C, the traffic carried by M trunks when offered 2 erlangs is given in Table 5.2. The number of additional tandem connections, M, required to carry the overflow traffic A_O is a function of the traffic already carried on the tandem route. If this existing traffic is A_T, the number of additional tandem circuits required, ΔT, is given by the next highest integer satisfying the equation [2]

$$\Delta T = \frac{A_O}{\left.\frac{\partial E(A_T, T)}{\partial T}\right|_{B=\text{const}}} \tag{5.3}$$

The partial differential in the denominator is the slope of the curve of Figure 5.3. From the approximations to the Erlang-B formula in Equation 4.10, it may be seen that the slope of the trunk traffic curve for constant grade of service is:

$$\left.\frac{\partial E(A_T, N)}{\partial N}\right|_{B=\text{const}} \simeq \begin{cases} 0{\cdot}85 & B = 1\% \\ 0{\cdot}78 & B = 0{\cdot}1\% \end{cases}$$

for $5 < A_T < 50$ erlangs

Table 5.2 also shows the relative costs for different proportions of high usage to overspill traffic. For instance, with a cost ratio of $\lambda = 1{\cdot}2$ the relative cost of all-direct to all-tandem routing is 6·0 to 3·6, so all-tandem is more economic than all-direct, as discussed in the previous section. However, the table also shows that an even cheaper solution is to provide two high usage trunks and one extra tandem trunk. This will carry the required traffic and the relative cost is only 3·2 — even lower than the all-tandem.

For a cost ratio of $\lambda = 2{\cdot}5$ the minimum cost again occurs for two high usage routes and the cost ratios of all-direct to optimum to all-tandem are 6·0 to 4·5 to 7·5.

The above calculations omit two factors:

(a) The cost of the extra complexity required in the local switching centres to perform automatic alternative routing. (For step-by-step systems this is an expensive addition but it is readily incorporated in common-control systems.)
(b) The different statistics of the overflow traffic (as discussed below). The effect of using more accurate calculations is to increase slightly the number of tandem circuits required to take the overspill traffic.

Characteristics of overspill traffic. Figure 5.5a shows a typical distribution of traffic during a busy hour. Figures 5.5b and 5.5c show the traffic carried if there

Table 5.2 Calculation of alternative routing strategy

No. of high usage trunks M	Traffic carried A_C (erlangs)	Traffic lost A_O (erlangs)	No. of tandem circuits* ΔN	Total relative cost			
				Cost ratios			
				1·2	1·5	2·0	2·5
0	0	2·00	3	3·6	4·5	6·0	7·5
1	0·67	1·33	2	3·4	4·0	5·0	6·0
2	1·20	0·80	1	3·2	3·5	4·0	4·5
3	1·58	0·42	1	4·2	4·5	5·0	5·5
4	1·81	0·19	1	5·2	5·5	6·0	6·5
5	1·93	0·07	0	5·0	5·0	5·0	5·0
6	1·98	0·02	0	6·0	6·0	6·0	6·0

*Tandem route assumed to carry traffic of 8 erlangs before overflow added.

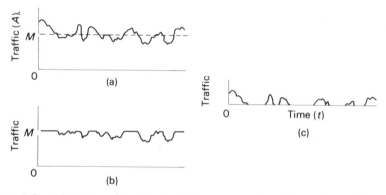

Figure 5.5 Character of traffic for high-usage route with overflow. (a) Traffic offered to high-usage route. (b) Traffic carried by high-usage route. (c) Overflow traffic.

are M direct trunks and the remainder carried on alternative routings. The high usage traffic is described by the Erlang-B formula. The traffic on the alternative route is very 'peaky' (more so than pure random traffic). The amount of peakiness is best described by the variance of the traffic. Pure random traffic with a mean of A erlangs is characterised by a Poisson distribution (Equation 4.3). The variance of this distribution is:

$$\mathrm{var}[P(k)] = \sum_{k=0}^{\infty} P(k)[k - A]^2 \qquad (5.4)$$
$$= A$$

The variance-to-mean ratio of a Poisson distribution is therefore unity. Overflow traffic is characterised by a distribution whose variance-to-mean ratio exceeds unity. It can be shown that, for a parcel of traffic, A_L erlangs, which results from the overflow of an offered load of A erlangs to M trunks, the distribution has a variance [3] given by:

$$V = A_L \left[1 - A_L + \frac{A}{1 + M + A_L - A} \right] \quad (5.5)$$

An overflow route generally takes traffic overflowing from a number of different routes. If there are n parcels of traffic with means of A_i erlangs and variances of V_i, the total traffic will have a mean, $A = \sum_{i=1}^{n} A_i$ erlangs and a variance of $V = \sum_{i=1}^{n} V_i$.

Wilkinson's equivalent random traffic technique. One method which is widely used to determine the number of trunks required in an overflow route is the equivalent random traffic technique devised by Wilkinson [4]. We have seen how to compute the mean and variance of the sum of the number of parcels of traffic on a particular route. Wilkinson's technique is to find the value, A^*, of the equivalent random traffic which when offered to M^* trunks results in an overflow with the same mean and variance as the actual traffic on an overflow route. Thus, the traffic originally described by the two parameters of A and V is described instead by two alternative parameters A^* and M^*. This equivalent traffic requires, say, N^* trunks to carry all the traffic at some specified grade of service, so the overflow requires $(N^* - M^*)$ trunks. This then is taken to be the number of trunks required on the overflow route.

It may be shown [5] that this equivalent random traffic is given (approximately) by:

$$A^* = V + \frac{3V}{A} \left(\frac{V}{A} - 1 \right) \quad (5.6)$$

and M^* by:

$$M^* = A^*/(q - A - 1) \quad (5.7)$$

where $q = 1 - 1/(A + V/A)$.

As an example, consider the case of 20 erlangs of traffic directed from B to C and 10 erlangs directed from B to A, as in Figure 5.6 (assuming all circuits are one-way). If there are 10 direct circuits from B to C, with 20 erlangs of offered traffic, the overspill traffic will be 10·8 erlangs and (from Equation 5.5), its variance will be 17·3. The mean and variance of the overflow plus other traffic on route BA are therefore 20·8 erlangs and 27·3 erlangs respectively. From Equation 5.6 the equivalent random traffic with this mean and variance is the same as

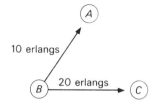

Figure 5.6 Example of alternate routing.

$A^* = 28·5$ erlangs overflowing from $M^* = 8·1$ trunks. (Note that this method uses non-integer numbers of trunks during the calculation.)

Now, 28·5 erlangs require 39·4 trunks for a 1% grade of service, so the number of trunks in the overflow route should be greater than $(39·4 - 8·1) = 31·3$, that is at least 32. Had we ignored the peakiness of the overflow traffic the result would have been that the total BA traffic of 20·8 erlangs would require only 31 trunks.

Choice of alternative routes. There are usually many possible alternative routings between two local centres. Figure 5.7 shows, as an example, the possible paths between two centres A and B, where each centre has a direct route to three other centres. When the direct route AB is busy, a routing may be attempted via, C, D or E. Now E has a direct route to B, but other routings exist such as EDB or ECB, or even ECDB. It is normally a design requirement of a switching centre that the choice of routes should depend upon only the destination. That is, the choice should be independent of the origin and of the route already taken. With this requirement, some care has to be taken with choice of alternative routes. For instance, D could be programmed to choose a route to E if the direct route to B is busy, and E could be programmed to choose a route to D if route EB is busy. These choices can lead to problems if routes DB and EB are both completely busy. In this case a call to B, arriving at D, will be routed to E. E finds EB busy, so it routes the call back to D, which then routes it to E, and so on. This is sometimes called the *ring around the rosie* problem [6].

This problem can be avoided if the signalling system between the centres can convey information about the routing already taken. Provision for this kind of information transfer is made in modern international signalling systems, where the additional cost of switching and signalling equipment is justified because of the resulting increased efficiency of use of the international trunks.

A more straightforward technique, for avoiding 'ring around the rosie' is to limit the choice of alternative routes at each centre. This is done by what is called the *far-to-near* strategy; when there are alternative intermediate centres, the first choice is the centre nearest to the destination centre. The final choice is the centre furthest from the destination centre, but nevertheless nearer than the centre making the choice. When this strategy is applied to the network of Figure 5.7, the first choice is the direct route AB. If this route is busy the first alternative is via the centre nearest to B, which in this case is E. E has a direct

124 TELEPHONE NETWORK ORGANISATION

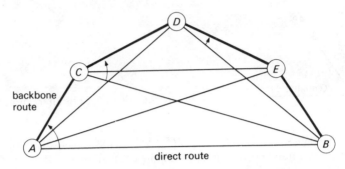

Figure 5.7 Choices of alternative routes.

route to B; if this is busy the call will be blocked, since routings via D or C are not allowed: they are further from B than E. If route AE is busy, the second alternative choice from A is AD. Now D can choose the direct route DB, or next DE, and so on.

If this strategy is applied to Figure 5.7, the valid routes between A and B, in order of choice, are:

AB	ACB
AEB	$ACEB$
ADB	$ACDB$
$ADEB$	$ACDEB$

There are three special types of route in this situation:

(a) A *fully-provided route*. The only route between two centres.
(b) A *high usage route*. The first choice route between two centres.
(c) A *final (or backbone) route*. The last choice route between two centres.

A routing consisting entirely of final routes connected end-to-end is called the *final routing* (or backbone routing) for the call in question. Note that all final routes must be fully provided, because calls which are offered to backbone routes will be blocked if not carried by those routes. However, by no means all of the fully provided routes are final routes.

Trunk networks. Long distance telecommunication is usually carried by separate networks from that of the local network. One reason is that different types of transmission multiplexing equipment are used for long distance communication and the functions of the switching centres are somewhat different (as is discussed below). Another reason, in some countries, is that local telephone service is provided by a number of different telephone companies, each serving a particular geographical area. These local networks are interconnected by a state-owned or state-regulated long distance network.

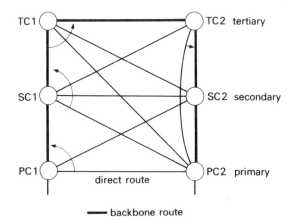

Figure 5.8 Principle of alternative routing in a hierarchical network.

As is described in Chapter 1, trunk networks are arranged with a hierarchy of switching centres, primary and secondary centres. In most countries the number of secondary centres is such that they have to be grouped together and parented by tertiary trunk centres [7]. In North America quaternary centres are also required [8]. At each level of the hierarchy the main routes are supplemented by large numbers of direct routes. Figure 5.8 shows some examples of possible routes. In addition, a local centre could be connected direct to any of the trunk centres, if this were justified by the volume of traffic.

Alternative routing in hierarchical networks. When alternative routing is applied to hierarchical networks an additional rule is required to avoid 'ring around the rosie'. The rule used is: 'if no direct route is available between a particular centre and a destination centre, this particular centre should first test (in far-to-near order) any possible routes via centres in the next hierarchical level above that of the destination.' There is no need to test *all* possible routes, but those that are tested must be in the order discussed.

Figure 5.8 shows how alternative routing is applied to a three-level trunk network in providing a route between two primary centres, PC1 and PC2. Each primary centre is parented on a different secondary centre. The secondary centres, SC1 and SC2 are patented on tertiary centres TC1 and TC2 respectively. TC1 and TC2 are directly connected by a trunk route.

Each centre has a list of alternative routes based on the rules described. For instance:

(a) *PC1*. The first choice is a direct route PC1–PC2, if it is provided and free. The next choices must be selected from the secondary centres. In this example there are only two, and these are selected in far-to-near order, that is, SC2, then SC1. There may be direct routes from tertiary centres, for terminating traffic but these are not normally used for originating traffic because they usually have high cost ratios. So, in this example, the final route from PC1 is that to SC1.

(b) *SC1*. By similar reasoning the choices from SC1 are PC2, SC2, TC2, and (final choice) TC1.
(c) *TC1*. This has three choices: PC1, SC2 and (final choice) TC2.
(d) *TC2*. This has two choices: direct to PC2 or (final choice) to SC2.
(e) *SC2*. This has only the final choice (fully provided) route to PC2.

It can be seen that strict adherence to these rules prevents the ring around the rosie situation occurring. In special circumstances these rules are relaxed as described below.

The route selection is performed link by link, so at first sight it appears that conditional selection would improve efficiency. For instance, if the direct route SC2 to PC2 is busy, but the route from SC1 to PC2 is not, and PC1 selects the route to SC2, the call will fail. However, had the later choice of PC1–SC1 been selected, the call would have succeeded. Nevertheless, various studies have shown that the increased gain in efficiency would be insignificant, and since the additional cost of the control and signalling to achieve conditional selection is high, the arrangement is therefore uneconomic.

In the hierarchical arrangement, the backbone routing between two primary centres consists of the routes:

$$PC \to SC \to TC \to TC \to SC \to PC$$

It guarantees an upper limit to the number of trunks connected end-to-end in a connection. In countries which have fully interconnected tertiary centres, the maximum number of links used end-to-end between local centres is seven. In North America, where quaternary centres are used, the maximum number is nine. However, the provision of direct and high usage routes implies that very few connections (typically less than 1 in 10^5 in the U.K.) actually use the backbone routing.

So, the advantages of the hierarchical network are:

(a) The traffic dispersal is near the economic optimum.
(b) No 'ring around the rosie' is possible.
(c) Routing decisions are made at each trunk exchange independently of the origin of the call or its routing so far.
(d) It provides an upper limit to the number of links connected end-to-end.

In practice the alternative routes do not always follow the plan. For instance, Figure 5.9 shows the actual routes (with circuit numbers) that are available between a particular switching centre in Chicago and one in New York City [9].

ROUTING PLANS   127

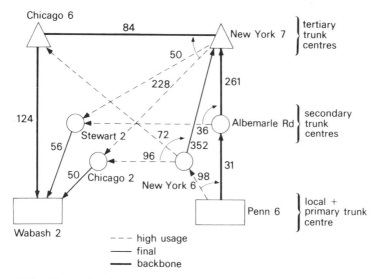

Figure 5.9 Example of practical alternative routing network. This diagram shows the possible alternative routes between two large No. 1 ESS local switching centres in New York and Chicago. These centres act as their own primary trunk centres. A call arising from Penn 6 to Wabash 2 attempts to use New York 6. New York 6 first choice is to Chicago 2 which has 50 circuits to Wabash 2. If all 98 circuits to New York 6 are busy, the call is set up via Albemarle Rd which attempts to set up the call via the Stewart 2 centre in Chicago. Final routing from Albemarle Rd is via New York 7 (a tertiary-level centre). New York 7 has 84 direct access to both Chicago 2 and Stewart 2 and a final group of 84 circuits to the Chicago 6 tertiary centre. This final group is sufficiently large to avoid the necessity for any further overflow to the quaternary centres in White Plains, New York and Norway, Illinois. (This diagram is adapted from MaCurdy, W. B. and A. E. Ritchie 1975. The network forging nationwide telephone links, *Bell Lab. Record*, **53**, 1, pp. 4–15.)

Use of four-wire switching. It is the purpose of the transmission plan to ensure that all permitted combinations of trunk routes provide an adequate quality of transmission [10]. Improved transmission results from four-wire switching wherever economically possible. In a modern trunk network all secondary and higher centres are four-wire switched. In some countries many of the primary trunk centres are also four-wire switched.

International routing. The international telephone network is based on a three-level hierarchy, as shown in Figure 5.10a. Centres of the lowest level of this hierarchy are called CT3s (CT stands for Centre du Transit). The CT3 is the interface between the national and international systems. There could be a number of CT3s in any one country. A CT3 may have a direct route to the destination CT3 or, failing that, it may be connected via a CT2 which serves a

128   TELEPHONE NETWORK ORGANISATION

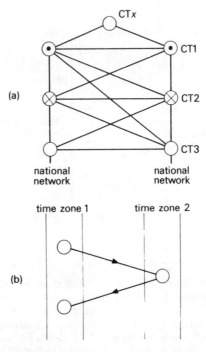

Figure 5.10 (a) The international routing plan. (b) Supplementary routing to utilise time differences between international exchanges.

defined part of the world. The highest level in the hierarchy is the CT1s, and seven CT1s are planned to cover the whole world. These are in (or planned to be in) London, New York, Sydney, Singapore, Moscow, Tokyo and somewhere in India or Pakistan. The CT1s are planned to be fully interconnected.

The basic routing plan is as described above, but the strictly hierarchical routing plan may be modified to take account of the special characteristics of the international network:

(a) The ratio of transmission to switching costs is considerably greater than a national network. The ratio of trunk to switching costs in a national network is approximately five, but for an international network it can exceed a hundred.
(b) With international circuits propagation times are significant.
(c) There are large time differences between the different countries, giving rise to strange traffic flow patterns, depending upon the relative times of day in the countries concerned.
(d) There are often commercial and political affiliations which impose restraints on certain routing patterns.

(e) It is highly desirable to avoid any routing containing more than one communications satellite circuit.

In order to cater for these situations, the CCITT permits modifications to the strict hierarchical system by allowing optional supplementary routings. Since these supplementary routings are optional, a transit centre which does not wish to use them does not have to provide the equipment to recognise and supply them.

One example of a supplementary routing is the interconnection of international centres via a transit centre which is placed in a different time zone, such as shown in Figure 5.10b. At certain times of day, there will be little traffic between the two time zones and the circuits will be under-used, so their use for this purpose increases their efficiency. (Similar routings are also used in North America where early morning traffic between East Coast centres is sometimes routed via the West Coast).

Another case where supplementary routing is desirable is when one of the links is already carried by a satellite circuit. In this case the call is best completed using land or submarine systems only. If there is no alternative, two satellite circuits connected end-to-end have to be used. In order to make this sort of decision the signalling system must be capable of indicating the route already taken.

Military systems [11]. Hierarchical routing is not generally applicable to military switching systems. These systems require a more dispersed type of routing to minimise the effect of the loss of a trunk route or a switching centre. This leads to routing plans based on hexagonal grids which can provide a wide range of alternative routes. A further feature of military systems is the mobility of some of the switching centres as they move with the group of people they serve. Some systems allocate a unique directory number to a subscriber. Since it is undesirable to provide within each centre a list of the location of every subscriber, a technique called saturation routing is often used. This involves transmitting a search message specifying the required number along all routes. A centre retransmits a search message for which it does not have the subscriber. The message which arrives first at the required location may be regarded as taking the shortest route and this is chosen for the connection.

5.3 Numbering plans

From the user's point of view it is desirable to provide a uniform numbering procedure so that it is possible to reach a given terminal from any other terminal by dialling the same code. This is called *Destination Coding*. Taking this concept to its logical conclusion means that every telephone terminal in the world should have a unique number by which it may be reached from anywhere else in the world. This is the long term aim of the CCITT. It involves using 14 or more digits. However, it is desirable to limit the number of digits dialled for three reasons:

(i) in many systems, especially step-by-step, the quantity of equipment provided is dependent on the number of digits dialled;
(ii) human factors — research indicates that the probability of dialling errors increases with the length of number; and
(iii) the longer the number, the longer it takes to dial, and this increases the holding time of registers.

So, it is usual, in public telephone networks, to have three lengths of number [12]:

(a) *Subscriber number* (or *customer number* in the U.S.A.) — a shortened version for local use.
(b) *National number* — standard length for use within the same country (or group of countries).
(c) *International number* — a full version for international use.

'Local', in the above context, usually means a particular community of interest. In practice this can be a small town and its outlying villages, a large conurbation such as London, or a complete state as in North America.

The provision of a uniform dialling procedure is thus a problem of fitting the machine to the man.

Subscriber number. This is the number to be dialled to reach a subscriber in the same local network or numbering area. In practice it is the number usually listed in the telephone directory and consists of up to seven, or possibly eight, digits at the most. Until the late 1950s these telephone numbers consisted of a combination of letters and numerals. This prevented the use of many codes, since they led to unpronounceable, inappropriate, or offensive combinations. Also it complicated the introduction of international subscriber dialling because there are differences between countries as to which letters correspond to which numerals. For instance, the letter 'O' corresponded to digit 0 in the U.K. but digit '6' in North America.

The subscriber number consists of two parts:

— exchange code,
— terminal number (or station number in North America).

In a multi-centre area the division between the two depends upon the size of the centres. An individual switching centre has anything from 100 to 100 000 subscribers and therefore requires from two to five digits for the terminal number. The remainder of the subscriber number specifies the particular centre. In general, the user is unaware of the split between the two parts.

National number. This is the number to be dialled within the same country or national numbering zone. For this purpose a country is divided into a number of separate *numbering plan areas*, each of which may be identifed by a unique series of digits called the trunk *code*. These area codes may be allocated geographically on a systematic or a random basis. If they are allocated on a systematic basis the code has some geographical significance. For instance, in the West German network the country is divided into seven zones, and the initial digit of the code indicates the zone. These are themselves subdivided into smaller areas which are indicated by the second digit. In a random allocation of codes the digits have no geographical significance. For instance, in the U.K. the code 21 is for Birmingham, 206 for Colchester, 225 for Bath.

If it is possible to devise a systematic set of codes, this simplifies some of the equipment designs and reduces the time taken to set up a call. However, it may be restrictive for future expansion of the number of terminals in a given area.

The trunk code plus local number is referred to as the *national (significant) number* [13]. In practice the number actually dialled for a national call includes a *trunk prefix* which distinguishes the call from a local one and which may be used to cause the equipment to give access to long distance call equipment. The prefix varies from country to country, but '0' is common in E.E.C. countries, although France uses '16'. North America uses '1' or '0', depending on whether it is a direct call or whether operator assistance is needed (for example, for a person-to-person or a transferred charge (collect) call). At the present time (1979) the trunk prefix is not always needed in North America. A national number dialled call is distinguished from a local call because the second digit of every trunk code is either 0 or 1, whereas no subscriber number has these digits in this position.

There is a limitation on the length of national numbers because they are used as part of international numbers. The CCITT has recommended that the number of digits dialled by a user for an international call should not exceed twelve and should preferably be only eleven [14]. This excludes any prefix needed to give access to the international service equipment. It will be seen in the next section that the country code may be one, two, or three digits long, and hence the national number should preferably be between 8 and 10 digits long (excluding any national number prefix).

The U.K. is divided into about five hundred areas which have been chosen on the basis of providing a convenient charging philosophy together with a minimum rearrangement of the pre-automation switching and line plant. The trunk codes consists of one digit for London, two digits each for five of the major cities, and three digits for the remainder. The subscriber number in London and the main cities is at present seven digits long, although London will probably go over to an eight digit scheme before the end of the century. The other areas of the U.K. have variable length subscriber numbers, but none exceeds six digits. Thus the maximum length of national numbers in the U.K. is nine. With the U.K. country code '44', this gives the recommended eleven digits for an international number.

North America [15], the United States, Bermuda and Caribbean Islands, Canada and a portion of North-West Mexico have been divided geographically into about 130 numbering plan areas (or n.p.a. for short). Most areas coincide with a state or provincial boundary, but the more populous states are divided into more than one n.p.a. For instance, California has eight, New York has seven and Ontario has five. The trunk codes here are all three digits and all subscriber numbers are seven digits long. The North American country code is '1' and this plan yields an international number of eleven digits for all terminals (unlike the U.K. which has between nine and eleven digits in its international numbers).

Other countries have different arrangements. Belgium, for instance, has been divided into ten regions, each of which contains one or more numbering plan areas. In the regions with more than one numbering area two digit codes are used, with the first digit peculiar to a region. So, the first digit of a Belgian area code has geographical significance.

International number. For international calls the world has been divided into nine *numbering zones* which are:

1 – North America	6 – South Pacific (Australasia)
2 – Africa	7 – U.S.S.R.
3) – Europe	8 – North Pacific (Eastern Asia)
4)	9 – Far East and Middle East
5 – South America	

An international number consists of an *international prefix*, to gain access to appropriate equipment, followed by the country code and the national (significant) number. The international prefix differs from country to country; in many E.E.C. countries it is '00', but there is less standardisation than with national prefixes. In the U.K. it is '010', in Belgium '91', in France '19', and in Sweden '009'.

As an example of these different dialling codes, consider the number Colchester 23456. Within the Colchester area it may be reached by dialling 23456 and ultimately it may be reached from anywhere in the U.K. by dialling 0206 23456, where '0' is the national number prefix and 206 is the area code for Colchester. Dialling from, say, Brussels it could be reached by 91 44 206 23456 where 91 is the Belgian international prefix, and 44 is the U.K. country code.

5.4 Register control of networks

The numbering plan is designed primarily for the convenience of the user of the system. However, an economic network is only achieved if a routing plan is

REGISTER CONTROL OF NETWORKS 133

chosen on the basis of traffic and costs. A switching centre has a number of routes which give it access to the other centres, and each route is identified by a route code. The centre must be able to translate a dialled number into a suitable routing.

For a local centre, the translation requirements are as follows:

(a) The local centre must store the dialled digits until sufficient digits have been received to determine the routing (for example, own centre, local or long distance). For a local call, up to three digits will generally be required to identify the terminating local centre.

(b) When the identity of the terminating local centre is determined, the originating local centre must translate the code into a suitable route and select and seize a free circuit on the chosen route. This route may be direct or via a tandem. When alternative routing is used, the translation process gives a result which is modified when a particular route is busy (that is all circuits in use).

(c) The process of translation and connecting the call to the required local centre generally takes longer than the inter-digit pause. If the user continues to dial the local number, before these operations are complete, the additional digits must be stored.

(d) When a direct route is available, the originating centre must transmit a *seize* signal to the terminating centre. When this is acknowledged, the originating centre must forward the required terminal number.

(e) When the selected route is via some other centre, this centre must be informed which of its outgoing routes is required. This may be achieved in two ways:
 — directly (for an 'unintelligent' centre) whereby the originating centre specifies the outgoing route. This is called *right-through routing;*
 — indirectly (for a more intelligent centre) whereby the originating centre simply repeats the trunk (or exchange) code to the intermediate centre which then performs its own translations to determine the outgoing route. This is called *destination code routing.*

The former method leads to a cheaper intermediate centre. However, it is more expensive to administer because any route changes out of this intermediate centre require modification to the translations of all the originating centres. The latter method is slower because it requires a translation process, but it is more flexible as it permits the intermediate centre itself to decide upon an alternative route, where necessary.

(f) A national number dialled call is (generally) recognised by the national number prefix (for example, '0'). The local centre must either:
 — connect the call immediately to the parent primary trunk centre and forward the national (significant) number as it is received;
 or
 — wait until sufficient digits have been received to determine whether a

direct routing exists outside the trunk network (for instance, two physically adjacent local centres parented on different primary trunk centres may in fact have a direct route even though they have different area codes).

The trunk network may be operated with right-through control from the originating primary trunk centre. To obtain the full benefit of alternative routing it is necessary to give some (if not all) of the higher rank trunk centres translation facilities. If they are so equipped, there are two ways in which they may be used:

(a) end-to-end control;
(b) link-by-link control.

In link-by-link control the originating trunk centre transfers the trunk code digits, plus subscriber number, to the intermediate trunk centre. This centre translates the code digits, establishes an outgoing path, and transmits the complete national number to the next intermediate centre, and so on. This technique implies that the intermediate centres must have registers capable of storing a complete national number. It also implies that the intermediate registers are held while the complete number is stored and subsequently pulsed out.

In end-to-end control the originating centre transmits only the trunk code to the next centre. This centre connects the call to a suitable outgoing route and then switches through the speech path. The next trunk centre in the connection transmits a signal back to the originating trunk centre requesting retransmission of the trunk code. This process continues for any number of trunk centres until the terminating trunk centre is reached. The terminating centre transmits back a special signal to the originating centre, requesting the local number (rather than the trunk code). End-to-end control implies that the registers in the intermediate centres need store only the trunk code (which is typically only three digits or fewer) and can have a significantly shorter holding time than with link-by-link control.

5.5 Use of step-by-step equipment

The discussion above shows how a flexible translation facility permits the complete separation of routing and numbering plans. However, many of the existing switching machines are based on Strowger-type step-by-step technology, where it is expensive to provide translation. Therefore many networks have evolved which involve non-ideal routings or a non-ideal numbering plan, in order to reduce the cost of the switching equipment.

Step-by-step switching centres may be directly connected together so that each digit (or digits), as they are dialled, directly operate the next switching

USE OF STEP-BY-STEP EQUIPMENT   135

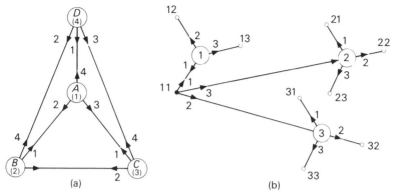

Figure 5.11   Use of direct control switching centres (numbers shown are dialled codes).

stage. This avoids the need for registers or senders. A small multi-centre network can be designed with a uniform numbering plan if the routing is either entirely mesh or entirely star. Figure 5.11a shows how the first digit dialled can be used to select a route to the required centre. This is known as a *linked numbering scheme*. The mesh method may be extended to a multi-tandem network, as shown in Figure 5.11b. Here the initial digit selects the appropriate tandem and the subsequent digits select the required local centre.

In the case of the star routing pattern, the outer centres act simply as concentrators. As soon as the calling terminal goes off-hook, a trunk is seized to the tandem; the first digit dialled selects the appropriate outgoing route. This technique implies that calls within a centre are still connected via the central tandem. (This is called a *trombone connection* after the shape of the instrument.) If this is a problem it is necessary to relax the objective of a uniform numbering plan and requires users connected to the outer centres to dial a prefix digit in order to gain access to the tandem. Thus, in Figure 5.11a the dialling codes are:

to	A	B	C	D
from A	1	2	3	4
B	91	2	93	94
C	91	92	3	94
D	91	92	93	4

An alternative solution to avoid trombone connections in star network is to provide a little 'intelligence' in the outer centres. When a calling terminal goes off-hook, a selector is seized in both the local centre and the tandem centre. Both respond to the first digit and, if the call is not for that centre, the selector is released.

It is possible to organise a trunk network on a strictly step-by-step principle if the country can conveniently be divided into up to ten regions and each region contains up to ten sub-regions, with 100 sub-sub-regions, and so on. With this network the primary trunk centre is accessed by the national number prefix and, from every trunk centre, there are up to ten routes to each of the ten tertiary centres covering the country. The next two digits identify the required secondary and primary centre respectively. This technique does not require any translation, but does imply that all the traffic for a region has to pass through the tertiary centre of that region. The addition of a simple translation facility at the primary centre permits direct routing to the required primary centre, or nearest secondary centre, if the routes exist. It also allows alternative routing.

Interworking of step-by-step with common control. A switching centre which has trunks incoming from a directly operated step-by-step centre has to be able to receive digits within about 100 ms of detecting a seize (that is within what is left of the minimum inter-digit pause after the i.d.p. has been detected by the last selector and that selector has selected a free circuit). With crossbar common-control machines, the register association time is generally at least 500–1000 ms, and longer under heavy traffic conditions. Under these circumstances it is necessary to do one of the following:

(a) provide special digit reception equipment permanently associated with each incoming trunk;
(b) stop the subscriber dialling, by telling him to wait for a second dial tone;
(c) provide a register/sender in the step-by-step system to store the dialled digits and repeat them when the common-control system has associated a register.

Solutions (a) and (c) are both in general use throughout the world. Solution (b) is sometimes used, especially on private networks. However, it introduces training problems for the user, especially if it is coupled with a non-uniform numbering scheme.

When step-by-step machines receive calls from common-control machines there is no problem, provided that one of the following occurs:

(a) the common-control machine can set up an outgoing route within an inter-digit pause;
(b) a second dial-tone is provided on the step-by-step machine;
(c) a register/sender is provided in the common-control system.

Solution (a) is possible with electronically controlled reed relay systems with dial pulse signalling. The same objections to (b) occur as above. Solution (c) is the most common solution because the common-control systems need to have a register anyway.

5.6 Charging

In a civilian telephone network it is necessary to charge users for the calls they make. The simplest method of charging is by a flat rate rental with all calls free, as this saves the cost of collecting any call data. This method is still used in many countries for local calls, but long distance calls are always charged on a time and distance basis. It is therefore necessary to collect the necessary information.

The information required for charging is the destination and duration of a call together with the identity of the calling party. For long distance calls it is generally uneconomic to provide the charging information at the local centre — this is normally done at the primary trunk centre to which the local centre is connected for inter-area calls. So, either the calling party's identity must be forwarded to the trunk centre or the charging information must be sent back to the originating centre.

The former method is standard in North America; after a local centre has transmitted the routing digits it follows them with a series of digits to identify the calling party [16]. This information is received at the trunk centre which transfers it to magnetic tape (or paper tape in older centres).

The charging equipment receives an *answer* signal from the called party and, at the end of the call, *clear* from either party. The time of answer and clear is transferred to magnetic tape, which may be processed later to produce the bill. This method has the advantage of providing an itemised bill, if the user wants one, and also the possibility of a large number of different tariffs for charging.

An alternative to itemised billing is what is called *bulk billing*; this is common in European countries. Rather than transmitting the line identity forward, the trunk centre sends back a simple *charge* signal to the local centre. This activates a meter connected to the calling terminal's line unit. The trunk centre determines the rate for the call and, on answer, transmits a single charge signal followed by a series of charge signals at a rate appropriate to the distance of the call. The maximum rate is about one every 0.5 second; this is determined by the response time of the mechanical counters, which are commonly used. A common method of generating the charge pulses is to derive them from a pulse stream at six times the required rate. The charge control unit sends a pulse on answer and then starts to count the charge pulses. After six have arrived it issues a new charge pulse. For the first period the counter ignores the first pulse.

Bulk billing is simpler than itemised billing, but market (or political) pressure for itemised billing exists. Bulk billing also is less accurate (to the advantage of the user) because the first time period may be up to 1/6 of a time unit longer. With itemised billing it is possible to time the call more accurately. It is also possible to charge a higher rate for the first minute, to take account of use of switching equipment.

5.7 Network management

A telecommunications network generally represents a very large investment. It is therefore important to ensure that it is always used in the most effective manner. The network must cope with unexpected peaks of demand and occasional failures of trunk routes and switching centres. The network design must ensure that the effects on the remainder of the system of local overload or failure are minimised. This is achieved by incorporating controls in individual centres and by providing network management centres which collect and process information from switching centres and transmission facilities in a geographical area. This information allows network controllers to detect the onset of overload or failures and to issue commands to contain the problem. This section describes the principles of network control and some of their applications [17].

There are two basic principles of network control:

(a) the more efficiently a network or switch is organised, the smaller margin it has for dealing with overload, and
(b) each network or switching centre has a maximum capacity; offering a load in excess of that limit may significantly *reduce* the capacity.

Effects of overload. There are two important performance criteria for a switching centre.

(a) Completion rate: the ratio between the number of call attempts, from users and incoming trunks, to the number of successful calls (defined as a connection to a local line or to a register in a distant switching centre). This is the performance as seen by the user. Under congestion conditions the ratio increases because users have to make repeat attempts when they discover congestion (or are timed-out while waiting for a resource, such as a local register or sender).

(b) Effective traffic: the total of conversation time which generates revenue for the telephone administration. The total traffic measured as trunk occupancy is generally higher than the effective traffic because this total includes set-up time and ineffective attempts. The ineffective time increases as congestion increases, for many reasons. The set-up time is increased as queues form for registers and senders and the ineffective time increases as calls progress further through the network, but fail at a late link in a multi-link connection.

The general relationship between offered traffic and effective traffic is shown in Figure 5.12. Below the design maximum the effective traffic is nearly equal to the offered traffic. Near the maximum there is some loss of traffic as is intended in the design. However, when the offered traffic exceeds the design maximum, there is a sharp decline, rather than a levelling off, in the effective traffic. A

NETWORK MANAGEMENT 139

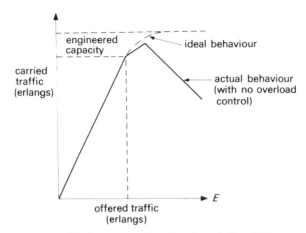

Figure 5.12 Relationship between offered and carried traffic.

further effect is that after a period of overload it takes some time for the system to recover. Both these effects are analogous to the well-known effects exhibited by road traffic systems. The traffic carried by a trunk road measured in vehicles per hour leaving the road is reduced in times of congestion. This arises because, once congestion occurs, a car slows down and spends a longer time using the resource (the road).

With direct routes a local traffic overload between centres affects only that route. However, with tandem and alternative routing, a local overload affects other routes as well. For a situation as shown in Figure 5.4 a local traffic overload between B and C will cause the traffic on the tandem links to increase and thereby affect any other traffic from B to A. With alternative routing the average number of links per connection increases when there is congestion, so the apparent carried traffic increases, while the effective traffic is reduced. Computer simulations of typical networks show that the more efficiently an alternative routing network is designed for normal load, the worse is performance is under overload. Under overload conditions it becomes necessary to cancel alternative routing to minimise these effects [18],[19].

Types of overload. There are three general classes of overload which require different control techniques:

(a) General overload. This affects a whole network and is caused by increased activity on a holiday or particularly busy days.

(b) Local overload. This is caused by a local problem, such as major equipment failure. The difficulty here is to prevent the effect of the congestion spreading into the rest of the network.

(c) Focused overload. This is generally caused by local disasters such as earthquakes, typhoons, or aeroplane crashes, where the calling rate from the rest of the network into this area becomes abnormally high.

All these types of overload have an effect on switching centres and trunk networks. In all cases controls must be introduced to limit the traffic at source, to prevent calls which have a high probability of being abandoned entering the system. Such calls may occupy resources for a significant time before being abandoned.

Overload within a (common control) switching machine is most easily detected by monitoring the number of calls which are waiting for a register. This is found to be a stable and sensitive indicator, since the queues build up quickly once the machine's maximum capacity is approached. This indicator also has the advantage that it is independent of the number of registers in service. When the number of calls waiting exceeds some preset limit, the machine is assumed to be in a state of near congestion and some form of traffic limitation is introduced.

For machine congestion caused by locally originated calls, the simple technique for traffic restriction is to disconnect temporarily the calling side of all but predetermined priority subscribers (such as police and doctors). A more sophisticated method is to split the non-priority subscribers into a number of groups and withhold calling detection from all but one group for several minutes. If the accepting group is rotated, the effective calling rate is reduced, assuming that all but the most persistent callers will be discouraged from making a call. This equipment may be added to common-control systems and is simple to incorporate in a computer controlled switching centre.

When the machine overload is caused by incoming trunks it is necessary to busy out a percentage of those incoming trunks. The signalling system must of course have the ability to transmit a backward busy signal, even for one-way trunks. In both the above cases of machine overload, the controls may be applied automatically or by manual intervention after the queue indicators have activated an alarm.

Overload in the trunk network is more difficult to detect. When a high percentage of the circuits on a particular route are busy, that route is congested. But in an alternative routing network congestion on the high usage routes is not in itself significant. When a route carries direct traffic plus overflow traffic, some protection is provided by permitting alternative routing only if the number of free circuits in the overflow route is above a preset minimum. This prevents the overflow traffic swamping the directly routed traffic.

Another technique used to improve the effective traffic capacity of a network is called *directionalisation.* When a machine recognises the onset of trunk congestion it switches some of its both-way trunks to the higher order centres to incoming only. This provides a preference for terminating, rather than originating traffic. Thus calls which have already passed through several links are given a greater probability of success compared with new calls, which have so far used fewer resources.

Network control centres. The techniques described above are successful in conditions of general overload and some cases of local overload. However, for focused overload, major equipment failures and other emergencies, these techniques are inadequate as they only utilise information local to a centre. To cater for more extreme situations it is necessary to provide traffic control centres which can receive telemetered information of all the major congestion indicators in a given geographical area. These network control centres are generally manned. The traffic supervisors detect congestion and issue commands to deal with them. Typically, the indicators are telemetered every 10–15 minutes to the control centre.

The commands may be to instruct major switching centres to modify their route translation codes to direct traffic around a major failure and to reject traffic focused onto an area unable to deal with it. Under these conditions emergency traffic is dealt with by operators who by the use of special codes are able to bypass these traffic rejection points. The commands may be carried out manually or, with computer controlled equipment, they may be carried out automatically. Most large countries with highly developed alternative routing networks are building such traffic control centres [20], [21].

References

1. For details of these accounting techniques see: De Garmo, E. P. and Canada, J. R. (1973) *Engineering Economy*. Macmillan Company.
   or
   Wherry, A. B. (1975) Investment appraisal in project planning, in *Telecommunications Networks* (Chapter 12). Peter Peregrinus.
2. For a detailed discussion of this approach see: Pratt, C. W. (1967) The concept of marginal overflow in alternative routing, *Aust. Telecoms. Res.*, 1, pp. 76–82.
3. This is derived in the classic paper on alternative routing by Wilkinson, R. I. (1956) Theories for toll traffic engineering in the USA, *Bell Syst. Tech. J.*, 35, pp. 421–514.
4. op. cit.
5. Rapp, Y. (1964) Planning of junction network in a multi-exchange area, *Ericsson Technics.*, 20, p. 77.
6. Clos, C. (1954) Automatic alternate routing of telephone traffic, *Bell Lab. Record.* 32, pp. 51–7.
7. For descriptions of the U.K. network see: Francis, H. E. (1959) The general plan for subscriber trunk dialling, *Post Office Elect. Eng. J.*, 51, pp. 256–67.
   Toblin, W. J. E. (1967) The signalling and switching aspects of the trunk transit network, *Post Office Elect. Eng. J.*, 60, p. 165.
   Wherry, A. B. *et al.* (1974) The London sector plan (four articles), *Post Office Elect. Eng. J.*, 67, p. 1.
   The German network is described in Hettwig, H. (1964) *Direct Distance Dialling*. Siemens & Halske.
8. The North American long distance network is described in: *Notes on Distance Dialling 1972*, A.T. &. T. Co.
   and
   Clarke, A. B. and Osborne, H. S. (1952) Automatic switching for nationwide telephone service, *Bell. Syst. Tech. J.*, 31, p. 851.
9. This example is taken from the special 50th Anniversary issue of *Bell Lab. Record*.
10. A full discussion of transmission planning is contained in: Hills, M. T. and Evans, B. G. (1973) *Transmission Systems*. Allen & Unwin.

11. A review is contained in:
    Cullyer, W. J. *et al.* (1975) Military communications networks, in *Telecommunications Networks* (Ch. 11). Peter Peregrinus.
12. *CCITT Orange Book* (1977) Definitions relating to national and international numbering plans, Vol. VI-1, Recommendation Q11.
13. *CCITT Orange Book* (1977) Numbering and dialling procedures for international service. Vol. VI-1, Recommendation Q10.
14. *CCITT Orange Book* (1977) Recommendation Q10 op. cit.
15. Nunn, W. (1952) Nationwide numbering plan, *AIEE Trans.*, 71, Part 1 pp. 257–60.
16. Dagnall, C. H. (1961) Automatic Number Identification, *Bell Lab. Record*, 39, p. 97.
17. Network management is very well covered in review paper: Gimpelson, L. A. (1974) Network management: design and control of communications networks, *Elect. Comm.*, 49, 1, pp. 4–22.
18. Burke, P. J. (1968) Automatic overload controls in a circuit switched communications network, *Proc. Nat. Electron. Conf.*, 24, pp. 667–72.
19. Laude, J. L. Local network management – overload administration of metropolitan switched networks, ibid pp. 673–678.
20. Averill, R. M. and Machol, R. E. (1976) A centralised network management system for the Bell system network, *International Switching Symposium*, paper 433–3.
21. Mumment, V. S. Network management and its implementation on the No. 4 ESS, ibid, paper 241–2.

6
Practical signalling systems

6.1 Types of Signal for Telephone Systems

The previous chapter indicated the need for a range of signals to be interchanged between switching centres. This chapter describes the signals that are needed and in particular describes the development of standardised international signalling systems. These have been one of the essential building blocks in providing the ultimate objective of a world-wide automatic telephone service. Standardised signalling provides the interface between different national systems.

The information that must be transmitted between switching centres falls broadly into three classes; supervisory, register and management.

Supervisory signals. These are the signals necessary to initiate a call set-up and to 'supervise' it once it has been established. For instance, over a trunk used for only one direction of call set-up, the signals needed from originating to terminating ends are:

– *seize*   – *release*

Those needed in the reverse direction are:

– *accept*   – *clear*
– *answer*

In addition certain administrative signals are often needed. For instance, the equipment at the receiving end may not always be able to receive traffic. It is therefore necessary to use reverse signals such as *block* and *unblock* to 'busy out' and 'unbusy' the trunk.

When the trunk is used for either direction of call set-up, the signalling system must be able to detect and resolve simultaneous seizures from both ends.

144   PRACTICAL SIGNALLING SYSTEMS

In a system which has 'calling party clear' or 'either party clear' operation, clear down must be prevented under certain circumstances. For instance, on a call to an operator, an emergency service, or special test equipment, the release of the call has to be under the control of the called equipment. In their cases a signal such as *inhibit clear* must be used. (In the U.K. this signal is called *manual hold* since it is normally extended from the operator's manual board.)

In some trunk control units it is necessary to provide what is called a *guard time* which keeps a trunk circuit marked as busy until the equipment in the remote centre has had time to return to its idle state: this is especially important in the Strowger-type systems. If this guard is not provided there is the possibility of a new seizure before the remote equipment becomes free and this leads to call failures. Because the guard time increases the ineffective holding time of the trunk circuit, it reduces its traffic capacity. So, a system which can signal when it is free by a *release guard* signal is preferable to one which uses a fixed guard time long enough to cover the worst case. The release guard signal must be separate from *clear* since the *clear* signal should be given as early as possible to allow release of the equipment before the trunk control unit.

On international calls it is desirable to provide the capability for an operator in the originating country to obtain access to an operator in the terminating country once a call has been established. This is necessary if the originating operator has some language difficulties once a call has been established. The signal to provide this facility is called *forward transfer*.

Routing information.   During the call set-up phase, a large amount of information transfer is needed to indicate the required number. Signals carrying this information are usually referred to as register signals. The basic information is the *routing* signal. This is the dialled code (or a translated version) which indicates to the subsequent switching centres the required routing. The *routing* signal normally consists of a series of decimal digits based on the numbering plan.

For operators' use in the international network, two further signals are used. These are referred to as 'code-11' and 'code-12'. These codes gain access to incoming operators at remote centres or to a specific operator used to book future international calls. In some networks 'code-13' is also used to gain access to special test equipment.

In addition to the basic routing information there is further information which can usefully be transmitted, especially in the international network. This additional information includes:

(a) Language digit.   On operator initiated international calls this digit precedes the routing information to indicate to the remote centre which language an operator should speak (if an operator is required). On a user-dialled call the digit is '0'.

TYPES OF SIGNAL FOR TELEPHONE SYSTEMS 145

(b) Route information. On international calls it is necessary to ensure, as far as possible, that not more than one link in the overall connection is a communications satellite circuit. Also it is necessary to indicate whether an echo suppressor will be needed at the remote end. This information also normally precedes the detailed routing information.

(c) Terminal information. Calls may be treated differently at remote centres depending upon the classification of the originating terminal. For instance, a call arising from an operator would have more facilities available than one originating from a normal user. Also, it is desirable to know whether a call arises from a pay 'phone. If the switching system uses centralised billing the full identity of the calling terminal is needed after the routing information has been passed. This information may also be needed in future systems to provide some of the more advanced facilities that are envisaged.

(d) Register control signals. In addition to the routing and other information, signals are also needed to control the registers themselves. These may consist of different initial signals, indicating the type of connection needed. In particular, a call may be a 'through' (transit) call or a call terminating at the centre in question (a terminal call).

With numbering plans which make it possible for the originating centre to know when a complete number has been received, an end-of-pulsing signal is added after the numerical information. This is referred to as an *'ST Signal'*.

All these register control signals are sent in the forward direction. As explained in the previous chapter, more control over the call is obtained if register signals in the backward direction are also available. In this case the following signals may be used.

(e) Acknowledgement signal. This indicates to the originating centre that the number has been received. Usually the acknowledgement signal also carries back information in the other categories below.

(f) Request for address. This can take many forms, depending upon the network organisation. In general, the request would be for either the required trunk code or the required subscriber number within the area. These signals are essential to provide fast set-up time in a system containing automatic alternative routing.

(g) Request for origin status. These signals can request the status of the originating caller (operator, payphone, priority user, etc.) or the full subscriber number.

(h) Congestion signals. These signals, if sent back, indicate congestion at a centre. They allow the originating centre to make a second attempt if it is able.

(i) Number complete. This signal confirms that all information needed has been received and, in the case of a register system, initiates the release of the register and the transfer of the system to speech conditions.

(j) Called terminal status. The signals included in this category include:

- Line free (with call charging);
- Line free (non-chargeable call);
- Line busy;
- Line out of order;
- Number not in use;
- Line on transfer.

The originating centre can then take appropriate action, for instance, provide tones locally on receipt of this information.

Once a sophisticated two-way signalling system is introduced there is a very wide range of control signals that can be used for the interchange. Examples of practical systems which provide all these signals are given later in this chapter.

Management signals. These signals are used to convey information or control between centres. This may involve remote switching of private circuits or temporary modification of routing plans, in the presence of overload. Increasingly, remote maintenance and change of terminal class of service are also being performed by means of signals. However, in most cases these signals are sent over separate data links, independent of the main transmission circuits.

Types of signalling equipment. As may be seen from the above discussion, two main types of signal are needed:

- Supervisory signals, which can occur at any time during the call.
- Routing signals, sent only while the call is being set up.

Since high speed set-up is necessary, in large systems sophisticated signalling equipment is needed for the routing information. However, for the supervisory information much simpler equipment is acceptable if only a limited range of signals is needed. For this reason it is often economic to provide the signalling equipment as two separate items of equipment. These are referred to as *line signalling equipment*, which is permanently associated with each circuit, and *register signalling equipment*, which is switched in as required. The sophistication of the line signalling equipment depends on whether register signalling equipment is available, and if so, whether backward register signalling is also provided. If two-way register signalling is provided, simpler line signalling is possible. For instance, if no backward register signalling path is provided, the line signalling equipment may be used to send busy and congestion signals.

The following section discusses examples of some of the signalling systems that have been developed. A continuing constraint on any new signalling system is that it should be able to interwork effectively with existing signalling systems, without imposing undue conversion cost when the signalling systems have to be changed. A new telephone switching centre which is installed in an existing area has to interwork with many different signalling systems on the various incoming and outgoing trunks. The need for compatibility often implies that all signalling systems convert the signals to some internal standard within the switching centre. Within the centre the signals may be sent from an incoming line to the selected outgoing line by d.c. conditions on a number of wires. One of the economic advantages of computer controlled switching for international use is that all the different signalling systems are interfaced to the computer system, which provides the appropriate conversions between them economically.

There would be considerable long-term advantage if a telephone network could be developed with a common end-to-end register signalling system, although the line signalling system could be performed on a link-by-link basis. The most convenient solution is a separate channel digital signalling system. However, this implies that all interworking exchanges are themselves computer controlled. In a network which develops over the years this is not the case and the introduction of separate channel signalling is involving high interface costs to the existing signalling systems at the interworking exchanges.

6.2 Signalling techniques

Line or supervisory signals. Line signals can be transmitted by the use of a single control channel in each direction, in parallel with the speech channel as shown conceptually in Figure 6.1a. These signal channels are referred to as the E and M leads. The condition presented to the M lead is reproduced at the E lead on the other terminal. Some of the techniques for provision of this control channel are discussed below but they are all used in a similar manner. Figure 6.1b shows one way that the sequencing of on/off conditions on each control channel can be used to provide a limited 'alphabet' of signals. Although these signals are simple to produce and detect, it is difficult to extend this alphabet, especially for a two-way trunk. Additional signals can be produced by introducing timing information, as shown in Figure 6.1c.

The actual method of providing the signalling channel depends upon the transmission technique used for the trunk circuit. There are basically three techniques of transmission:

(a) Physical pairs. Each trunk is provided by a single pair of wires (with amplifiers if necessary) and there is a d.c. path between the two ends. In this case signalling from the originating to terminating end of the trunk can be performed by the loop condition at the originating end (i.e. open or closed). Signalling in

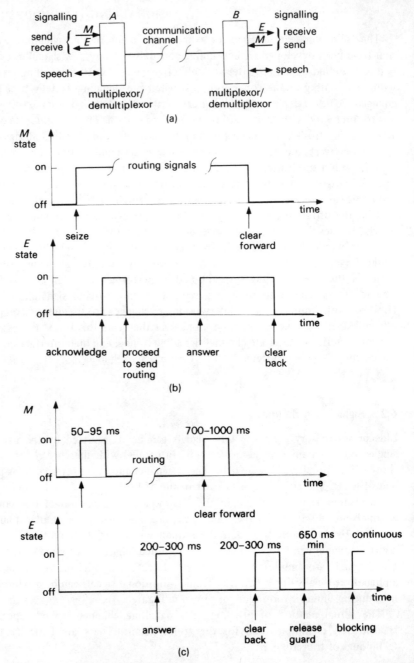

Figure 6.1 Use of simple line signalling system. (a) Addition of signalling channels to speech channels. (b) Example of use of *E* and *M* leads. (c) Pulsed use of *E* and *M* leads. (Note that there is no explicit *proceed to send* in this system. Also that *clearback* and *answer* are unambiguous as they must occur alternately. This system may be used on both-way trunks.)

the reverse direction can be achieved by changing the polarity of the power feed to the trunk (as shown in Figure 6.2a). This technique can also be used for two-way trunks, in which case, in the idle condition, each end is looking for a loop at the other end.

(b) Frequency division multiplex (f.d.m.). For long distance trunk routes with 12 or more circuits it is more economic to combine a number of speech signals into a composite signal with each speech channel occupying a different frequency band. The basic principle is shown in Figure 6.2b. Because a two-wire circuit carries speech signals in both the forward and backward directions, the two directions must be separated before multiplexing is carried out. This separation is achieved by a device called a hybrid transformer. The speech from the line is then modulated using single sideband (suppressed carrier) modulation and added to 11 other signals similarly modulated with different carriers. The carrier frequencies of each channel are spaced at 4 kHz intervals from 60 to 108 kHz and the filters limit the speech energy to roughly between 300 and 3,400 Hz. The 12 channels can then be transmitted as one. This multiplexing process may be repeated; a number of 12-channel groups may themselves be modulated with different carrier frequencies and combined to make a super-group. For large capacity systems this process may be repeated several times. The maximum size of the final system is at present a collection of 10 800 channels occupying a bandwidth of about 60 MHz.

Line signalling information can be transmitted over f.d.m. systems by sending a sinusoid tone, which is controlled by control lead M, over the voice channel. This tone may be *in-band*, that is within the band 300 to 3400 Hz (typically in the range 2280 or 2600 Hz) [1]. Alternatively it may be *out-band* (typically 3850 Hz) [2]. Figure 6.2b shows how the signals are added and removed in the two cases. In-band signalling is simpler but has the drawback that there is a possibility of speech signals imitating a control signal. This problem is greatest once a call is established since the line receiver is connected across the line looking for a *clear* signal. An in-band signalling receiver therefore normally has a *guard* circuit which inhibits its operation if there is energy outside the signalling band. In other words, to produce a reliable *clear* signal there must be energy at the signalling frequency and not anywhere else.

A second problem with in-band signalling occurs when several transmission links are connected end-to-end. The in-band signal may '*spill-over*' from one link to the next and cause false operation of another signalling system. For this reason, whenever a signal is detected, a line split relay is operated to disconnect a link from the succeeding one and thus prevent spill-over. Typically this line split relay is designed to operate within 35 ms [3] and this figure therefore defines the minimum recognition time for signals in subsequent links.

Out-band signalling is more complex but does not suffer from speech imitation. It is also possible to use out-band for signalling during the speech phase without disturbing the conversation.

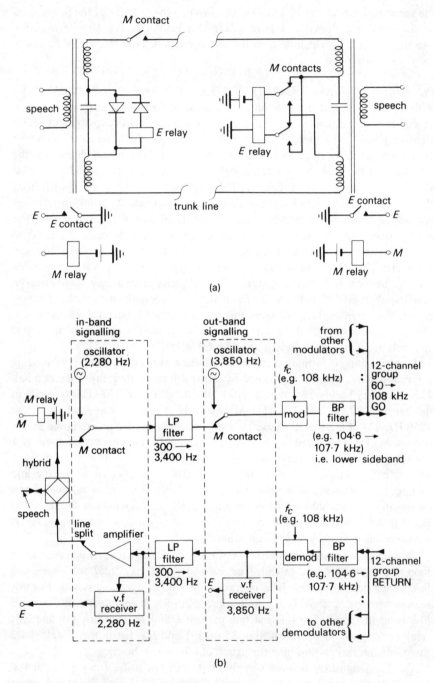

Figure 6.2 (a) Example of adding line signals to speech channel. (b) Principle of in- and out-band signalling on FDM systems.

(c) Pulse code modulation multiplex (p.c.m.). In p.c.m. multiplex systems, speech is converted by an analogue-to-digital converter to 7 or 8 bit samples at a sampling rate of typically 8000 samples/second. There are two systems in common use, one with 24 and one with 30 speech channels. On the earlier version of the 24-channel system the format of each channel consists of 7 speech bits plus 1 signalling bit [4]. A more recent version of the 24-channel system is used in North America and improves the quality of the speech signal by using the eighth bit from only one frame in six. This is called a multi-frame. For a sampling rate of 8 kHz this provides a signalling channel of capacity 1·33 kbits/second. This signalling channel is further subdivided into two separate channels of capacity 667 bits/second. The p.c.m. system itself provides a four-wire system and there is therefore a signalling channel in either direction.

The simplest method of use of the signalling bits is E and M signalling whereby the signalling bit in each direction merely transmits the loop condition in the forward direction and the battery polarity in the backward direction. Numerical signals are transmitted at 10 impulses a second and the sampling rate, even with a multiframe, is more than sufficient to transmit this accurately.

In Europe, the standard p.c.m. multiplex system is based on 30 speech channels transmitted within a frame of 32 time slots (numbered 0 to 31) [5]. This requires a total bit rate of 2048 kbits/second. Channel 0 is used for providing the framing signal and channel 16 is used for transmitting the signalling information relating to speech channels 1 to 15 and 17 to 31.

The frames of 32 time slots are grouped into a multi-frame of 16 frames (numbered 0 to 15). Each frame within the multi-frame transmits 4 bits of information for two of the channels. The net rate for each of the four signalling channels per speech channel is therefore 500 bits/second. (The 4 bits of frame 0 are used for supervisory information such as indicating loss of multi-frame alignment at the remote end.)

Alternatively the 64 kbits/second channel may be used as a separate channel signalling system such as described in the next section.

Register signals. The simplest way to transmit the register signals is to use loop-disconnect signals and transmit them with the line signalling system. This method is widely used, but it is slow and has an alphabet limited to the ten decimal digits.

A more rapid and flexible transmission technique which is in wide use is that of multifrequency (or m.f.) signalling. There are a number of such systems in use throughout the world, but they all employ combinations of two in-band frequencies selected from five or six possible frequencies. This gives an alphabet of 10 or 15 respectively. Note that multifrequency signals for trunk signalling use *all* possible combinations of five or six tones, rather than one tone from one group of four and one from another, as with signalling between the terminal and the switching system. Also, the frequencies used are different in the two cases.

152 PRACTICAL SIGNALLING SYSTEMS

Table 6.1 R1 signalling system (N. American system)

Line signals

Single tone in each direction in-band

Signal	Signal direction[1]	Duration[4]	Transmitted state[2,3] Originating end	Terminating end
Idle	⇌	continuous	0	0
Seize[5]	----→	continuous	1	0
Delay-dialling	←----	continuous	1	1
Proceed to send	←—	continuous	1	0
Answer	←---	continuous	1	1
Clear back	←—	continuous	1	0
Disconnect (clear forward)	—→	continuous	0	0 or 1
Busy	←—		0	1
Guard		automatic fixed period of 750 to 1250 ms after initiation of disconnect		
Release after a double seize	—→	on for 100 ms off 100 ms then on continuous		

1. —→ on, ----→ off.
2. 0 is tone on or signalling bit = 0 in p.c.m. system.
3. 1 is tone off or signalling bit = 1 in p.c.m. system.
4. Normal persistence time on or off is 30–40 ms, disconnect is 300 ms.
5. Double seize assumed to have occurred if no proceed to send is received within 5 seconds of seize.

Register signalling

(two tones selected from six — forward direction only) (link-by-link)

Combination number	Signal	Timing (gaps 70 ms)
1	Digit 1	
:	:	70 ms
9	Digit 9	
10	Digit 0	
11	KP (Start of pulsing)	100 ms
12	ST (end of pulsing)	70 ms
13–15	Spare	

The signalling system developed by Bell Laboratories for use between switching systems within North America (an international version of which is called the R1 system by the CCITT) uses six frequencies[6]. For national use, 12 of the 15

SIGNALLING TECHNIQUES 153

possible combinations are used; the two signals extra to the ten decimal digits are used to indicate the start and end of a sequence of digits. Reliable signal detection is possible with 60 ms pulses and 80 ms interdigit pauses, so a ten digit routing code can be transmitted in 12 x 120 ms, that is in less than 1·5 seconds. In the international R1 system, all 15 possible digit signals are used. The line and register signals used are shown in Table 6.1.

A more recent system, called R2, has been defined by the CCITT (Table 6.2) [7]. This provides two-way high capacity signalling and is intended for use both nationally and internationally. In R2 the frequencies are spaced 120 Hz apart and are in the ranges:

$$1380 \text{ to } 1980 \text{ Hz for forward signals, and}$$
$$540 \text{ to } 1140 \text{ Hz for backward signals.}$$

The backward signals allow a much wider repertoire of control signals. R2 contains a 'shift' capability (analogous to shift on a typewriter to change character case) to increase the signal repertoire further.

Table 6.2 R2 signalling system

Line signals

		Analogue		Digital			
				forward		backward	
Signal	Direction	forward	backward	a_f	b_f^1	a_b	b_b
Idle	$\rightleftarrows$	on	on	1	0	1	0
Seize[2]	$\rightarrow$	off	on	0	0	1	0
Seize acknowledge	$\leftarrow$	not provided		0	0	1	1
Answer	$\leftarrow$	off	off	0	0	0	1
Clear back	$\leftarrow$	off	on	0	0	1	1
Clear forward	$\rightarrow$	on	on or off	1	0	0	1
Block	$\leftarrow$	on	off	1	0	1	1
Release guard	$\leftarrow$	on	on	1	0	1	0

Line signals — link-by-link
  Analogue version uses single out-band tone in each direction (20 ms recognition time).
  Digital version uses two bits/channel of p.c.m. multiplex (20 ms recognition time).

Notes
1. b_f is used to signal a fault at originating end.
2. For detection of double seize and subsequent actions see Reference [7].

154 PRACTICAL SIGNALLING SYSTEMS

Register signals

	FORWARD		BACKWARD	
Signal No.	GROUP I address information etc.[1]	GROUP II Nature of call (reply to A-3, A-5)	GROUP A Acknowledgement to Group I	GROUP B State of subscribers line
1	Digit 1 (French)	Subscriber	Send next digit (n + 1)	Sub. line free (last party release)[2]
2	Digit 2 (English)	Priority	Send last but one digit (n − 1)	Sub. transferred
3	Digit 3 (German)	Maintenance equipment	Address complete / Changeover to B signals	Sub. line busy
4	Digit 4 (Russian)	Spare	Congestion (national)	Congestion
5	Digit 5 (Spanish)	Operator	Send calling class of service	Unassigned number
6	Digit 6	Data transmission	Set-up speech path (charge)	Sub. line free − charge
7	Digit 7	Subscriber	Send last but two digits (n − 2)	Sub. line free − no charge
8	Digit 8	Data transmission	Send last but three digits (n − 3)	Out of order
9	Digit 9	Priority sub.		spare for national use
10	Digit 0 (i.s.d. call)	Operator		
11	Code-11	Coin-box[2]		
12	Code-12 (request not accepted)			spare for international use
13	Automatic test call			
14	Incoming half echo suppressor required			
15	ST (End of pulsing)			

(Group II column bracket: NATIONAL / INTERNATIONAL)

Table 6.2 (continued)

Register signals consist of 2/6 tones in each direction. The signals are sent end-to-end with compelled signalling as follows:

- on seizure outgoing register starts to send first signal;
- when incoming register recognises signal, it responds with backward signals which act as an acknowledgement as well as having their own meaning;
- as soon as outgoing register recognises backward signal it terminates outgoing signal;
- as soon as incoming register recognises cessation of signal it stops sending backward signal;
- as soon as outgoing register recognises cessation of acknowledgement, it may, if necessary, start to transmit next signal.

Notes

1 Alternative meanings when used as initial digit are shown in brackets.
2 Assigned for national use.

Common channel signalling. In a switched network a significant proportion of the total cost is involved in applying analogue signals to one end of a trunk circuit and then removing and decoding them at the other end. It is in fact the most appropriate technique when the control of a switching system is distributed over a large number of sub-systems. However, an increasing number of telephone switching centres use computer control. When one computer controlled system has trunks to another such system, it is more economic to send all the control information direct from one computer to another via a data link. This is known as *common channel signalling*. Figure 6.3 shows the format of a system of this type designed for international signalling called the CCITT No. 6 signalling system [8]. It uses a 2·4 kbits/second data link and one link can deal with up to 2048 trunks. Line and register information relating to a particular trunk is transmitted in one or more of the 28-bit packets (20 information bits plus 8 bits of error-checking information). The first packet relating to a new signal for a particular trunk contains an indication of the number of packets and an identifier for the particular trunk.

A similar example is the system being adopted by the Bell System for the North American network; information is transferred by signal units consisting of 28 bits of which 20 are for data and 8 for error checking [9]. Information may be transmitted either in a lone signal unit (l.s.u.) or as a multiple unit message of up to six signal units. The format of an initial (or lone) signal unit is shown in Figure 6.4a. The first bit indicates whether it is an initial or subsequent signal unit. In the initial signal there are 11 bits to indicate which trunk circuit is involved (i.e. up to 2048 trunks) and 8 bits are available for information. Subsequent signal units of a multiple unit have the format shown in Figure 6.4b together with typical uses of the information.

156 PRACTICAL SIGNALLING SYSTEMS

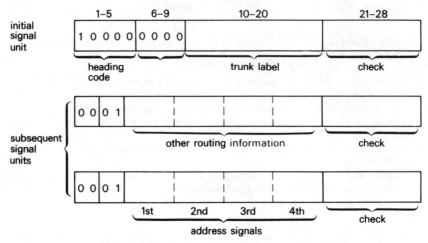

Figure 6.3 Format for CCITT No. 6 separate channel signalling system.

At the receiving end of a link each signal unit is checked individually for errors and a single acknowledgement signal unit (a.s.u.) (Fig. 6.4c) is transmitted after the reception of every eleven signal units.

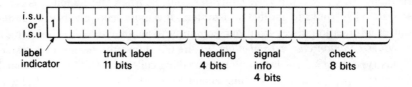

	no. of signals			binary codes	
	available	assigned	spare	heading	signal info.
i.s.u. of i.a.m.[1]	16	1	15	1000	0000 to 1111
i.s.u. of s.a.m.[2]	7	7	0	1001 to 1111	0000
s.a.m. as l.s.u.	105	77	28	1001 to 1111	0001 to 1111
l.s.u.[3]	120	34	86	0000 to 0111	0001 to 1111
i.s.u.[3]	8	0	8	0000 to 0111	0000

(a)

SIGNALLING TECHNIQUES 157

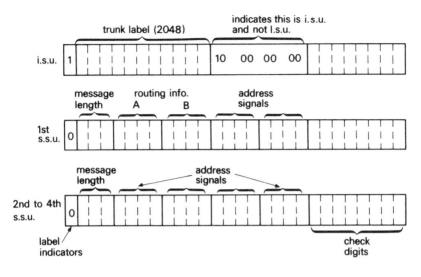

field	no. of signals			example of use
	available	assigned	spare	
message length	8	6	none available	no. of s.s.u.s
routing info A	4	3	1	terminating/transit satellite used, echo suppressor used spare
routing info B	16	12	4	language digit, call and customer category
address signals	16	14	2	decimal digits, code-11, code-12, ST

(b)

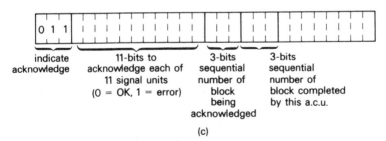

(c)

Figure 6.4 Formats for Bell CCIS (taken from Reference 9). (a) Initial or lone signal unit formats (i.s.u. or l.s.u.). (b) Initial address message (i.a.m.). (c) Acknowledgement signal unit (a.s.u.) sent after each group of clear signal units. (Note: (1) initial address message, (2) subsequent address message, (3) non-address type signals (such as network management).)

158   PRACTICAL SIGNALLING SYSTEMS

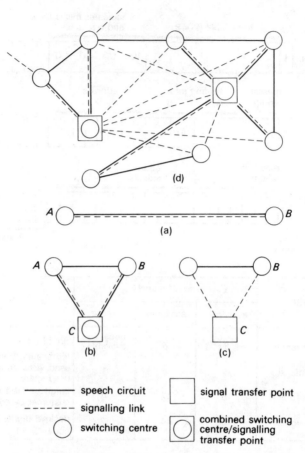

Figure 6.5   Separate channel signalling system (CCITT No. 6). (a) Associated mode. (b)–(c) Quasi associated modes. (d) Double star mode of quasi associated.

Since the signalling is separate from the speech, it is not necessary to transmit it by the same route. Figure 6.5 shows two types of signal routing:

(a)   *associated mode*;
(b)   *non-associated mode* where the signals are transferred between two centres over one or more common signalling links in tandem. The signals are processed and forwarded through one or more *signal transfer points*. The non-associated paths may be *fully disassociated* where each signal transfer point acts as a switch or *quasi associated* where the transfer is over predetermined paths only.

The relative economics of signal links and speech links imply that the non-associated mode is more attractive. In this mode all centres in an area are connected to only two signal transfer points (to provide network redundancy) in the form of two stars (Fig. 6.5d) even though there may be direct links of the speech paths [10].

One advantage of an associated channel signalling system is that it automatically checks the transmission path. In order to ensure satisfactory performance in a separate channel signalling system it is necessary to perform a continuity check of the speech circuit: the originating end applies a tone of 2100 Hz and the presence of this tone is checked at the receiving end.

The data link itself uses a 2·4 kbits/second data link. The signal units are assembled into a continuous bit stream and synchronising patterns are added. If no information is being sent, a suitable null pattern is transmitted.

Since one signalling system may be responsible for between 1000 and 2000 trunk lines, a very high degree of system security is required. This involves duplication of signalling paths between centres.

A new version of CCITT No. 6 is being developed for use on p.c.m. multiplex systems. This is known as CCITT No. 7.

References

1. For typical systems see either Battista, R. N. *et al.* (1970) A new single-frequency signalling system, *Bell Lab. Record*, 48, 3, pp. 85–9.
   Miles, J. V. and Kebon, D. (1962) Signalling system AC No. 9, *Post Office Elect. Eng. J.*, 55, pp. 51–8.
2. *CCITT Orange Book* (1976) Systems recommended for out-band signalling, Vol. IV–1, Recommendation Q21.
3. *CCITT Orange Book* (1976) Splitting arrangements and signal recognition times on 'in-band' signalling systems, Vol. VI–1, Recommendation Q25.
4. *CCITT Orange Book* (1976) Characteristics of primary p.c.m. multiplex equipment operating at 1,544 kbit/s, Vol. VI-1, Recommendation Q47.
   For a typical description see
   Gaunt, W. B. and Evans, J. B. (1972) The D3 channel bank, Bell Lab. Record 50, pp. 229–233.
5. *CCITT Orange Book* (1976) Characteristics of primary p.c.m. multiplex equipment operating at 2,048 kbit/s, Vol. VI-1, Recommendation Q46.
6. *CCITT Orange Book* (1976) Signalling system R1, Vol. V1-3, Recommendations Q311–332.
7. *CCITT Orange Book* (1976) Signalling system R2, Vol. V1-3, Recommendations Q350–368.
8. *CCITT Orange Book* (1976) Specification of signalling system No. 6, Vol. IV-3, Recommendations Q251-Q267.
9. Bell Systems version is described in Dahlbom, C. A. (1972) Common channel signalling – a new flexible interoffice signalling technique, *International Switching Symposium*, pp. 421–7.
10. Nance, R. C. and Kaskey, B. (1976) Initial implementation of common channel interoffice signalling, *International Switching Symposium*, paper 413–2.
    See also special issue of *Bell Syst. Tech. J.* (1978) 51, No. 2, on Common Channel Interoffice Signalling.

7
Design of switching networks

7.1 Basic multi-stage networks

In Chapter 3 it was explained that the use of a common switch network is one of the techniques for achieving economy in a switching centre. There are two types of network required. Figure 7.1a shows a N-inlet N-outlet network needed for switching the speech path and Figure 7.1b shows an N-inlet M-outlet ($M < N$) network used as an access switch. The requirements for an access switch differ from those for a speech switch. In a speech switch a particular inlet must be connected to a particular outlet whereas in an access switch a particular inlet may be connected to any one of a number of free outlets.

A straightforward way to connect N inlets to N outlets is the square matrix such as shown in Figure 7.10b which uses N^2 crosspoints. This arrangement is represented more concisely in the form shown in Figure 7.2a, or even more simply, as in Figure 7.2b–d. In this chapter notation (c) is used. Notation (d) is used in diagrams in later chapters where it is only required to convey general information about the number of stages and their interconnection of devices.

In this chapter it is assumed that the one aim of a switch network designer is to minimise the total number of crosspoints required to switch a given traffic load at a given grade of service. In practice, however, it is not the only aim, other factors to be taken into account include complexity of control, expandability and overload performance. The approach for time-division switching is completely different and is discussed in Chapter 12.

Instead of using individual crosspoint devices such as relays, a square network could also be built from a number of uniselectors, with their outlets connected together as shown in Figure 7.3. In telephone switching centres this parallel connection of outlets is called *multiplying* and the group of wires providing the connections is called a *multiple* (as the wires are connected to a multiplicity of switch outlets).

BASIC MULTI-STAGE NETWORKS 161

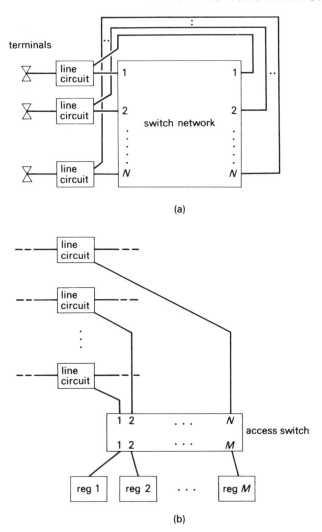

Figure 7.1 Types of switch network. (a) Basic network – N inlets to N outlets. (b) Access switch – N inlets to M ($< N$) outlets.

Because crosspoints are provided to connect a terminal to itself and to connect terminal B to terminal A as well as terminal A to terminal B, some crosspoints are redundant. A more economical arrangement called a *folded* or *triangular matrix* is shown in Figure 7.4. This uses only $\frac{1}{2}N(N-1)$ crosspoints rather than N^2.

In both the square and folded matrices a connection between two free terminals can always be made, irrespective of other connections in the network. In most switching systems this is an unnecessarily costly arrangement because, in

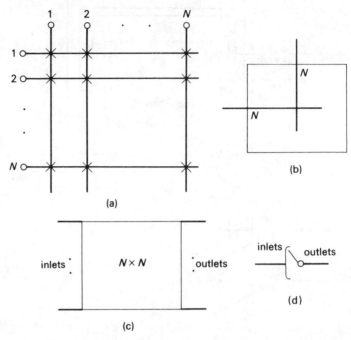

Figure 7.2 Representations of a switch matrix.

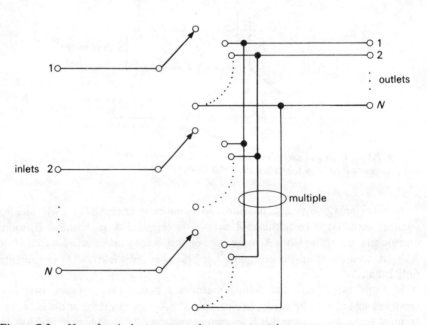

Figure 7.3 Use of uniselectors to make square matrix.

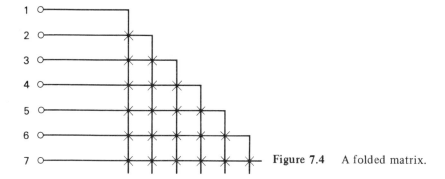

Figure 7.4 A folded matrix.

general, terminals are used for only some proportion of the time; these simple networks provide unnecessary switching capacity.

Two-stage networks. The examples above are of single stage networks — only one crosspoint need be operated in order to connect an inlet to an outlet. It is possible to reduce the total number of crosspoints if the switch network is constructed from a number of stages consisting of smaller matrices. The method used is illustrated in Figure 7.5a with a 10 x 10 square matrix as the basic building block. One matrix can connect any of 10 inlets to 10 outlets. If each of these 10 outlets is connected to 10 different matrices, each with 10 outlets, the original inlets have access to a total of 100 outlets. Two crosspoints have then to be operated to make a connection.

This two-stage system introduces a new problem, that of *internal blocking*. This is defined as the probability that a free inlet fails to be connected to a free outlet because of the absence of a path. Internal blocking occurs in the two-stage network because there is only one link between a particular first-stage switch and each of the second-stage switches. A connection between any pair of first- and second-stage switches precludes any further access between that pair. Ways of minimising the effect of this internal blocking are discussed later.

In the arrangement of Figure 7.5a each of the second-stage switches has nine inlets unused. These other inlets can therefore be used to receive links from nine other first-stage switches as shown in Figure 7.5b. This arrangement provides a network with the same number of inlets as outlets.

In order to make a larger network a further stage of switching must be added as shown in Figure 7.5c. First ten 100 x 100 sub-networks are provided and the 100 outlets from these sub-networks are connected to 100 separate third-stage switches. These then provide access to a total of 1000 outlets. The extra inlets on the third-stage switches can now be used by nine other 100 x 100 sub-networks, as shown in Figure 7.5c.

Continuing this arrangement for K stages, each stage consisting of $n \times n$ matrices, gives a network with N inlets and N outlets, where $N = n^K$. Figure 7.5d

164 DESIGN OF SWITCHING NETWORKS

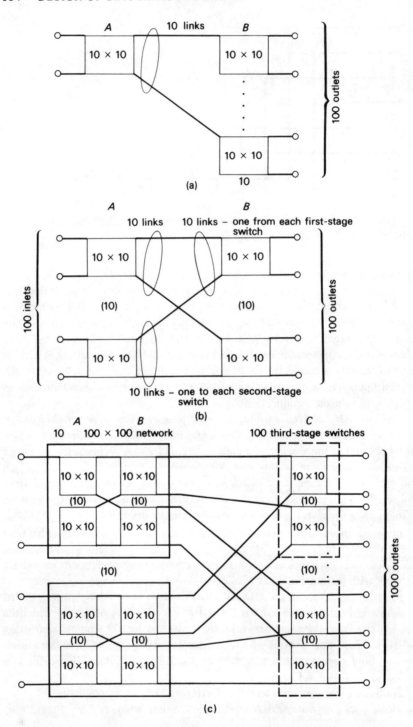

BASIC MULTI-STAGE NETWORKS   165

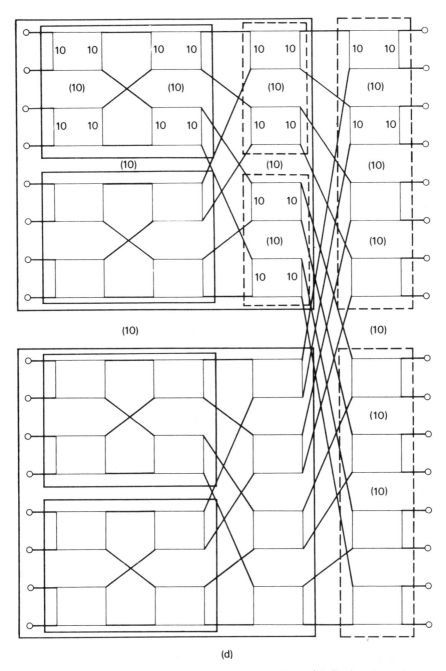

(d)

Figure 7.5 Construction of multi-stage networks. (a) Basic two-stage network. (b) Increase in number of inlets. (c) Three-stage network. (d) Example of a four-stage network.

shows a four-stage network. In this general arrangement each stage of switches is called a *distribution stage* because outlets are distributed over an increased number of stages, and each stage increases the number of reachable outlets by a factor n. Note that each stage of switches consists of the same number of matrices, N/n.

Optimum size of component matrices. With K stages of $n \times n$ matrices suitably connected, the number of outlets is $N = n^K$. In other words, for N inlets and N outlets, using an $n \times n$ matrix the number of stages needed is:

$$K = \ln N / \ln n \qquad (7.1)$$

The total number of crosspoints in the K-stage network, X, can be shown to be:

$$X = Kn^{K+1} = nN \ln N / \ln n$$

If the only objective is to minimise the total number of crosspoints the optimum value of n from the above expression can be found to be $n = e(= 2 \cdot 718 \ldots)$. Since n must be an integer, the optimum size of matrix is therefore 3 x 3.

However, applying this result to a 1000-inlet 1000-outlet network, the optimum number of stages will be $K = \ln 1000 / \ln 3 = 6 \cdot 28$. Taking $K = 7$ and $n = 3$ yields a 2187-inlet 2187-outlet network using a total of 45 927 crosspoints or 21 crosspoints per inlet. By comparison the three-stage 10 x 10 network provides 1000 inlets and uses 30 000 crosspoints, that is 30 crosspoints per inlet.

The above analysis ignores the problem of internal blocking. When this is taken into account the optimal arrangement is based on a larger matrix (see below). Also, the cost of controlling a multi-stage network increases with the number of stages. If this is taken into account, the use of even larger component matrices may be found to be more economic.

Internal blocking. With an increasing number of distribution stages, the probability of internal blocking increases. This probability may be estimated by the following simple argument:

We shall assume that the traffic per inlet is a erlangs and therefore the proportion of time that each inlet is in use is a. Because there are as many internal links as inlets, the probability that each link is busy is also a (assuming that the probability of an inlet wanting a connection to a particular outlet is the same for all outlets).

In a multi-stage network, of the type described above, there is only one path between a specific inlet and a specific outlet and this path consists of $K - 1$ links connected end-to-end. Assuming that there is no correlation between the link occupancies in different stages (and this assumption has been found sufficiently adequate in most practical systems), the probability of one link being free is

$(1-a)$. The probability of all $K-1$ links being free at the same time is $(1-a)^{K-1}$. Therefore the blocking probability, B is given by

$$B = 1 - (1-a)^{K-1} \qquad (7.2)$$

Table 7.1 shows the blocking probabilities for different sized networks for average occupancies of 5% and 10% per terminal ($a = 0.05$ and 0.1). From this it is clear that, for most practical purposes, the simple multi-stage network is inadequate. It is necessary to provide more than one path between each first and last stage switch in order to reduce the probability of blocking. The way this can be done is described in the next section.

Table 7.1 Blocking probabilities for simple multi-stage networks

No. of stages	B (a = 0·05) (%)	B (a = 0·1) (%)
2	5·0	10·0
3	9·7	19·0
4	14·2	27·1
5	18·5	34·4
6	22·6	40·9
7	26·4	46·9
8	30·1	52·2

7.2 Use of mixing stages

Figure 7.6 shows how the number of paths between a given inlet and outlet can be increased. The first two stages of the network consist of $n \times n$ matrices, so any inlet on the first stage has access to n^2 links from the second stage. In order to provide more paths, these n^2 links are connected to only n third-stage switches (rather than n^2, as in a distribution stage). This type of link pattern is called a *mixing stage*. It can be seen from Figure 7.6 that the addition of a mixing stage does not increase the number of outlets accessible from a first-stage switch, but it does increase the number of paths between switches on the first and last stages.

There are now n possible paths, each consisting of two links connected end-to-end, so internal blocking will occur only if all n are busy, so that:

$$B = [1 - (1-a)^2]^n \qquad (7.3)$$

For a three-stage 10×10 network, with $a = 0.1$, this gives $B = 6 \times 10^{-8}$ (which is negligible).

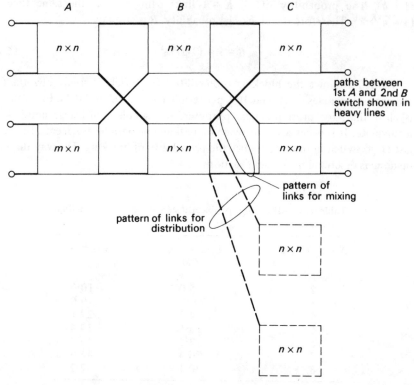

Figure 7.6 Addition of mixing stage to two-stage network.

Appendix C shows, that when blocking is taken into account, the optimum size of switching stage is:

$$n_{opt} = e/(1-a) \quad \text{for} \quad a \leqslant 1 - \frac{e}{N} \quad \text{or}$$

$$n_{opt} = N \quad \text{for} \quad a \geqslant 1 - \frac{e}{N}$$

rather than $n = e$, and the optimum number of stages is given by

$$K_{opt} = \frac{\ln N(1-a)}{\ln n(1-a)}$$

where n is the nearest integer to n_{opt}.

For a network with $a = 0\cdot 1$ erlang, the optimal network will have $n = 3$. In this case a 1000-inlet network requires at least seven stages. This gives at least 205 crosspoints per erlang. For $N = 10\,000$ at least 10 stages will be needed (for $n = 3$) giving at least 275 crosspoints per erlang.

Appendix C also shows that the minimum number of crosspoints per erlang is achieved when N is large and $a = 0.5$ erlang. This implies that $n_{opt} = 2e$ (~5·436) and the minimum number of crosspoints per erlang is then given by

$$E > 4e \ln \tfrac{1}{2}N$$

For $a = 0.5$ erlang, the best practical value of n is 5 and for $N = 1000$ this gives 67·6 crosspoints per erlang.

Non-blocking networks. It is possible to build a multi-stage network in which no internal blocking occurs and which still uses fewer crosspoints than a square array. The solution to this problem was first published by C. Clos [1] in 1954. An example of a network capable of switching any of 100 inlets to any of 100 outlets without blocking is shown in Figure 7.7. The 100 inlets and 100 outlets are both served by 10 switches, each of which has access to 19 links (The reason for the figure 19 will become apparent later.)

It can be shown that this network is non-blocking by considering the worst possible case for setting up a connection. This occurs when a connection is required between an inlet on one of the A-stage matrices which already has all the other 9 inlets in use. Because there are 19 links from this matrix, there are still 10 free links to the B-stage matrices. In the worst possible case the required outlet is served by a C-stage matrix on which there are already 9 other outlets in use. However, there are still 10 free links to this matrix from the B-stage matrices. Since there are only 19 B-stage matrices there must be at least one B-stage matrix which is served by a free link from the A stage and a free link from the C stage. It follows from the above argument that it is always possible to make a connection through this network between a free inlet and a free outlet.

This particular network uses a total of 5700 crosspoints, as compared with the 10000 in a simple square array; the saving in crosspoints is not very great.

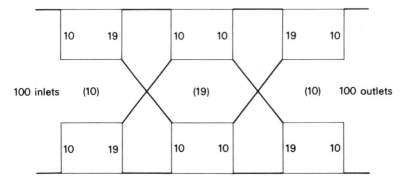

Figure 7.7 Three-stage non-blocking (Clos) network.

The fact that Equation 7.3 is only an approximate derivation of B is clear in this case because it gives a small finite value for B, whereas B for a non-blocking network is, of course, zero.

It may be seen that, in general, solution of a three-stage non-blocking network has

n $n \times (2n - 1)$ matrices in the A and C stage
$2n - 1$ $n \times n$ matrices in the B stage

Other non-blocking solutions are possible by the provision of further intermediate stages. For networks above a certain size this provision of extra stages may reduce further the total number of crosspoints.

7.3 Network and channel graphs

A four-stage network consisting of three distribution stages and one mixing stage is shown in Figure 7.8a. In the diagram only the first and last links to and from each switch are shown. Because it is the pattern of links which is important in determining the overall blocking probability, it is convenient to draw the links in the form shown in Figure 7.8b which is known as the *network graph*. Each switch is represented by a dot and each link by a line joining appropriate dots in each stage. This diagram shows only the general pattern of links so if there are $n \times n$ switches there will in fact be n links from each dot to n separate dots in the next stage.

An even simpler graph can be drawn for a particular inlet and a particular outlet. Only the links and intermediate stages which could be used between the particular inlet and outlet are shown. This graph is shown in heavy lines in Figures 7.8a and (b) and is drawn separately in Figure 7.8c. (Note that it is the pattern (or topology) that is important. The exact position of the dots is irrelevant.) This graph is very important in the design and analysis of multi-stage networks and is called the *channel graph* of the network. Most of the networks treated in this book are symmetrical and therefore the channel graph has the same form between each of the inlets and each of the outlets.

The channel graph of K distribution stages is a series of $K - 1$ single links because between a particular inlet and a particular outlet there is only one path.

The four-stage network shown in Figure 7.8 is not the only possible method of arranging a mixing stage. The network and channel graphs of the two other methods are shown in Figure 7.9. These differ in the way in which their mixing stages are arranged. Each mixing stage 'collects' links which have originated from earlier stages. For instance, in Figure 7.8a the mixing stage is the final one, and the links collected have diverged from the first stage.

A convenient notation for these arrangements has been developed by Takagi [2]. Distribution stages are denoted by D and mixing stages by M. A relationship line is used to denote the stage from which a mixing stage collects its paths. So

NETWORK AND CHANNEL GRAPHS 171

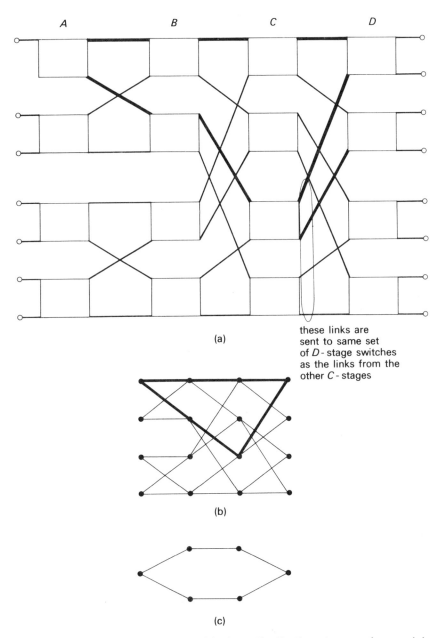

Figure 7.8 Four-stage network with three distribution stages and one mixing stage. (a) Basic network. (b) Network graph. (c) Channel graph.

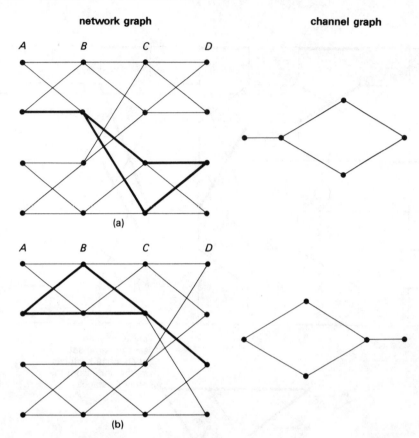

Figure 7.9 Alternative networks to Figure 7.12. (a) Mixing stage at D – collecting links diverging from B. (b) Mixing stage at C – collecting links diverging from A.

Figure 7.8a is $\widetilde{DDDM}$ whereas Figure 7.9a is $\widetilde{DDDM}$. Figure 7.9b is $\widetilde{DDMD}$ and it can be seen that this is the mirror image of 7.9a. When a channel graph is reversed, so that the right and left sides are interchanged then (if X represents either M or D)

$$\widetilde{D\ldots X} \text{ becomes } \widetilde{X\ldots M}$$
$$\widetilde{X\ldots M} \text{ becomes } \widetilde{D\ldots X}$$
$$\widetilde{X_1 \ldots M \ldots X_2} \text{ becomes } \widetilde{X_2 \ldots M \ldots X_1}$$

and the relationship lines are maintained between the same stages, that is

$$\widetilde{DDMDM} \text{ becomes } \widetilde{DDDMM}$$
$$\widetilde{DDMDM} \text{ becomes } \widetilde{DDMDM} \text{ it is unchanged because}$$

the network is symmetrical)

NETWORK AND CHANNEL GRAPHS

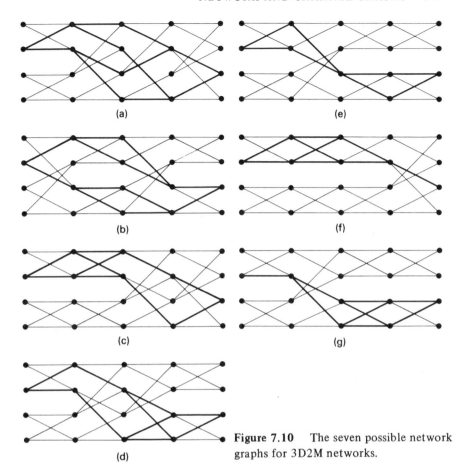

Figure 7.10 The seven possible network graphs for 3D2M networks.

A useful exercise is to find the channel graphs corresponding to the different ways of arranging three distribution stages and two mixing stages. (There are seven ways and they are shown in Figure 7.10.)

Takagi optimal channel graphs. By a rather complex argument Takagi [2] has shown that, for a given number of distribution and mixing stages, there exists a particular pattern of interconnections which minimises the probability of internal blocking. This optimum is derived from two results as follows

Theorem 1
A network of the form

174   DESIGN OF SWITCHING NETWORKS

where Xs are either D or M stages (or absent), with mutually intersecting relationship lines, is superior in terms of blocking probability to one of the form

$$X\overbrace{DXDXMXM}X$$

in which the relation lines are parallel. An example is shown in Figure 7.11a.

Theorem 2
A network of the form

$$X\overbrace{DXDXM}X$$

is superior to

$$XDX\overbrace{DXM}X$$

An example of this is shown in Figure 7.11b.

Appendix D gives a simplified proof of these most important results and provides a numerical example to demonstrate their validity.

It is now possible to rank the seven graphs of Figure 7.10. From Theorems 1 and 2 it can be shown that they have relative merit

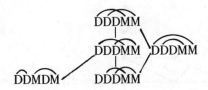

The optimal graph for K D-stages and l M-stages ($K \geq l$) is found to be of the form

$$\underbrace{D \ldots D}_{l} \underbrace{D \ldots D}_{K-l} \underbrace{M \ldots M}_{l}$$

In other words, the optimal graph is obtained by providing as many stages of distribution as are needed to give access to the required number of outlets and adding as many mixing stages as are needed to bring the blocking probability to below the design objective.

Some properties of optimal graphs are as follows:

They may be expressed in terms of just K and l.
They are symmetrical.
Optimal $2K$ graphs are not necessarily obtained by joining two K graphs together.

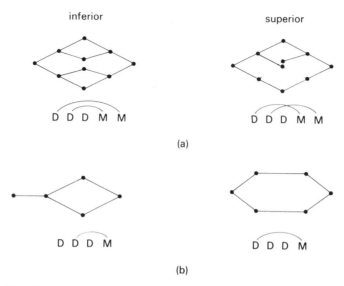

Figure 7.11 Relative merits of different channel graphs. (a) Theorem 1; (b) theorem 2.

In a later paper Takagi [3] extends his proof to cases where non-square component matrices are used. However, his theories still give no guidance on the relative merits of different numbers of stages or actual size of matrices to be used. Practical considerations of system growth often make it impractical to use the optimal graph and some compromises have to be made.

7.4 Networks with concentration

The previous sections dealt with networks built from square matrices. For networks with relatively low terminal occupancy (say $a < 0.1$) a higher crosspoint efficiency may be obtained if non-square matrices are used. A simple example is a network with N inlets and N outlets which must be able to connect up to M concurrent calls. This can be built by a two-stage network where an $N \times M$ switch connects the inlets to M internal links and a second $M \times N$ switch connects these links to the N outlets.

Figure 7.12a shows an example of this network for 1000 inlets able to connect up to 100 concurrent calls. This network uses 200 000 crosspoints. The blocking probability can be computed directly from the appropriate equations of Chapter 4. In this case the number of terminals can be assumed to be effectively infinite. If blocked calls are assumed to be lost, the Erlang-B Equation 4.7 will apply. The approximate form of Equation 4.11 tells us that 100 trunks will carry about 81 erlangs at 1% blocking probability and 72 erlangs at 0.1%. This corresponds to an average traffic per terminal of either 0.08 or 0.07 erlang.

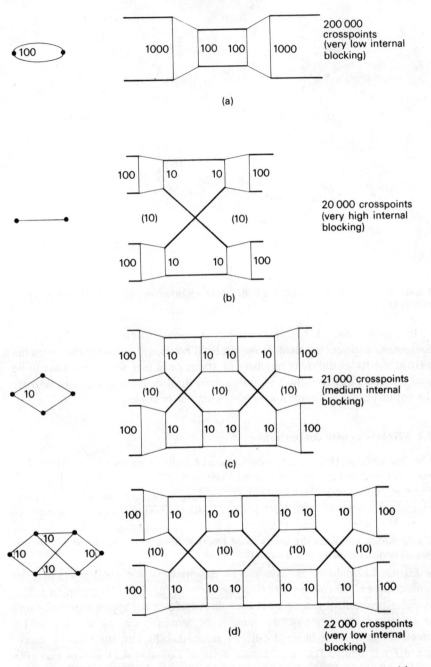

Figure 7.12 1000 inlet system with concentration. (a) Two stages with one matrix per stage; (b) two stages with ten matrices per stage; (c) three stage; (d) four stage.

NETWORKS WITH CONCENTRATION 177

The number of crosspoints required can be reduced to 20 000 if ten 100 x 10 matrices are used instead of the 1000 x 100 matrices. These smaller matrices must be interconnected as shown in Figure 7.12b to provide access from every inlet to every outlet. The occupancy of the internal links is (to a first approximation) ten times that of the inlets. So the probability of internal blocking is very high because there is only one link between a particular first and last stage matrix.

Figure 7.12c shows how a mixing stage can be introduced to reduce the internal blocking. The internal blocking is given approximately by Equation 7.3 with $n = 10$. This gives the following values:

a	$B(\%)$
0·3	0·1
0·4	1·2
0·5	5·6
0·6	17·5
0·7	38·9

Therefore the maximum practical link occupancy is between 0·3 and 0·4 erlang (or 0·03 to 0·04 erlang per terminal) and 21 000 crosspoints are needed.

In order to increase the traffic per terminal, a further mixing stage is needed. This is shown in Figure 7.12d which has 22 000 crosspoints. The blocking probability of this network is given (again approximately) by equation (D.5), with $X = a$. This can be computed term by term to give

a	$B(\%)$
0·4	0·012
0·5	0·34
0·6	1·02
0·7	4·32
0·8	16·6
0·9	53·3

So, the network of Figure 7.12d can carry roughly 0·5 to 0·6 erlang per link, that is 0·05 to 0·06 erlang per terminal at between 0·3 to 1·0 per cent blocking.

Note that these formulae are only approximate and are useful only as a guide to a network capacity. In a practical design it is usually necessary to perform detailed simulations in order to determine the actual traffic capacity of the network.

The above results are summarised in Table 7.2, where the significant improvement obtained in crosspoint efficiency can be seen. The best case is still far from a theoretical optimum; greater efficiency can be obtained if more stages of smaller sized matrices are used. Figure 7.13 shows a 10 000-inlet, 10 000-outlet network with low internal blocking using 100 x 10 and 10 x 10 matrices. Note

178 DESIGN OF SWITCHING NETWORKS

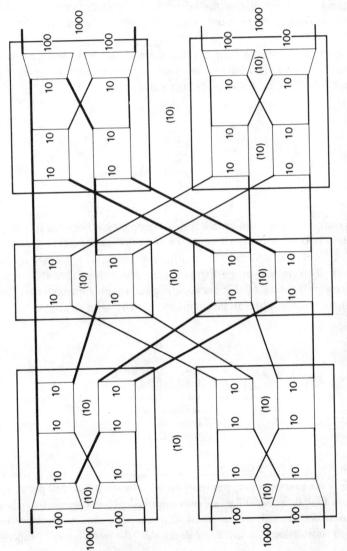

Figure 7.13 10 000 inlet network with optimum channel graph.

Table 7.2 Comparison of networks of Figure 7.12

Network	Traffic/terminal for about 1% internal blocking	Number of crosspoints	Crosspoints/erlang
Figure 7.12a	0·08	200 000	2500
Figure 7.12b	0·01	20 000	2000
Figure 7.12c	0·04	21 000	525
Figure 7.12d	0·06	22 000	366

that these diagrams now show that the sub-networks are in fact the line and trunk switch networks which were introduced in Chapter 3.

Practical results indicate that a high crosspoint efficiency depends on the terminal having access to about five crosspoints and two successive stages of concentrations are preferable to the one in Figure 7.12. These two stages of concentration bring the traffic per link to about the optimum of 0·5 erlang per link for greatest efficiency of the trunk switch network.

7.5 Lee's simulation technique for evaluating blocking probabilities

The channel graph provides a very convenient basis for evaluating the blocking probability of complex networks. The technique was originally developed by C. Y. Lee of Bell Laboratories [4]. The basic assumption is that the probability of a link being busy is independent of the state of all other links. This is the implicit assumption used in the earlier part of this chapter. The practical effect is to overestimate slightly the blocking probability.

This assumption can be used as the basis of a computer program [5]. A model of a channel graph is built within the computer memory and the average occupancy of each link is computed for a particular level of offered traffic. The program works by conducting a number of trials. It generates a series of random numbers and uses these numbers to decide whether a link is busy or free for a particular trial. For instance, if a link has an average occupancy of 0·2, and the random numbers generated are in the range 0 to 1, the program will set the link busy if the random number is below 0·2. This setting of state at the links is repeated for all the links in the channel graph; using a different random number for each link. Once all the links have been set busy or free, the program checks whether there is a path between the two end points. A large number of these trials are made and the program computes the proportion of the trials for which no path is found. This is taken to be the estimate of the blocking probability for that particular level of offered traffic. The accuracy of the result improves with the number of trials. The smaller the blocking probability the greater will be the required number of trials to achieve the same degree of statistical confidence. For instance, a run of 80 000 trials which yielded 80 blocked cases can be shown to give blocking probability of 0·1% with 95% probability that the actual value lies between 0·08% and 0·12%.

Application of Lee's hypothesis in effect says that the probability distribution of the number of concurrent calls on a set of links is given by the binomial distribution. This assumption is sufficiently accurate for most of the sets of links between stages. However, improved accuracy is obtained if the busy states of the first and last sets of links are chosen on the basis of the Engset distribution, as these links are affected by a very limited number of traffic sources [6].

References

1. Clos, C. (1953) A study of non-blocking switching networks, *Bell Syst. Tech. J.*, **32**, pp. 406–24.
2. Takagi, K. (1969) Design of multi-stage link systems with optimum channel graph. *Rev. Elect. Comm. Lab., Japan*, **17**, 10, pp. 1205–26.
3. Takagi, K. (1971) Optimum channel system of link system and switching network Design, Rev. Elect. Comm. Lab., Japan, **20**, pp. 962–86.
4. Lee, C. Y. (1955) Analysis of switching networks, *B.S.T.J.*, **34**, pp. 1287–315.
5. Grantges, R. F. and Sinowitz (1964) NEASIM. A general-purpose computer simulation program for the load-loss analysis of multi-stage central office switching networks, *B.S.T.J.*, **44**, pp. 965–1004.
6. Takagi, K. and Itoh, M. (1970) Internal blocking probabilities of eight-stage link systems for electronic switching systems, *Rev. Elect. Comm. Lab.*, **18**, pp. 840–53.

8
Control unit design

8.1 Role of control units

Previous chapters show that a switching centre consists of a large number of control units of different types such as line units, main control units and switch control units. Any control unit receives control signals from terminals or from other control units. The function of a control unit is to act on these signals, by operating devices such as switch networks and issuing signals to other control units.

This chapter examines techniques for specifying the requirements of control units and techniques for conveying signals between control units. In particular it discusses ways in which control units and signalling techniques can be selected to help maintain the availability of a switching centre even though individual control units may on occasions be faulty.

The techniques were introduced in Section 1.4 in the form of two types of diagram:

>signal exchange diagram;
>state transition diagram.

Both types of diagram can be used to specify the behaviour of different control units in a switching centre. The former specifies the 'alphabet' of signals that are passed between one control unit and another. The state transition diagram (s.t.d.) specifies the response of a control unit to any sequence of events (an event being the arrival of a signal). The s.t.d. is a powerful design tool; it forces a designer of a control unit to ask himself for every state, 'have I considered every possible event that could occur when the control system is in this state? What should be the reaction of the unit to each event?'

However, use of the s.t.d. assumes that events are treated sequentially. It does not indicate the required response when two or more events occur simultaneously. In any system of interconnected control units responding to random demands for service it is possible for a control unit to receive signals simul-

182 CONTROL UNIT DESIGN

taneously from more than one source. One example can be seen by consideration of the Basic Switching System shown in Figure 8.1. This consists of N calling control units connected to N called control units. Clearly, in the absence

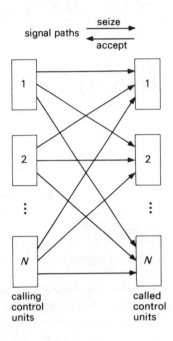

Figure 8.1 Signal paths between N calling and N called control units.

of any constraint, it is possible for more than one calling unit to transmit simultaneous *seize* signals to the same called control unit. Similar situations are likely to occur in a resource sharing system where several control units can make simultaneous attempts to seize the same resource.

In any real-time switching system it is necessary to provide arbitration so that a set of simultaneous signals arriving at a control unit can be ordered and dealt with, one by one. Section 8.2 introduces the general concept of an arbiter. Specific implementations of such units are described in Chapter 9.

8.2 The arbiter

Arbitration in a called control unit can be shown explicitly, by partitioning the called control unit into an arbiter unit and a main control unit (Figure 8.2). The arbiter has inputs from the N calling control units and can be in one of two states, *busy* or *free*. The operation of the arbiter may be specified as follows:

(a) If the arbiter is in the *free* state, a *seize* signal from any one of its inputs puts it into the *busy* state. The arbiter then sends a *seize* signal to its

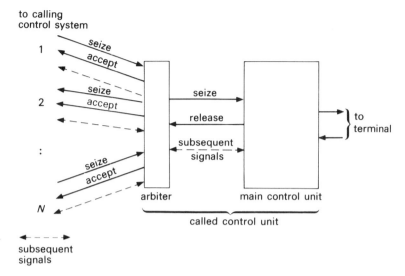

Figure 8.2 Partitioning of called control units into arbiter plus main control units.

associated main control unit and returns an *accept* signal to the calling control unit which has seized it.

(b) If the arbiter is in the *busy* state, a further *seize* signal on any of its inputs has no effect. (The calling unit detects that the called unit is *busy* either by the absence of an *accept* signal or by the return of an explicit *reject* signal.)

(c) If two or more *seize* signals are received simultaneously by an arbiter, only one will be accepted.

(d) Once seized, the arbiter acts as a switch and permits the exchange of subsequent signals between the calling and associated called control units.

(e) The arbiter remains in the busy state until it receives a *release* signal from the main control unit.

(f) The arbiter may be designed to perform a queueing function. In this case, when the arbiter receives the *release* signal from its main control unit, it acts on a previously received *seize* signal from another calling control unit. The arbiter may do this on a preset priority order or at random. However, if the order of service is to be a function of time of arrival (for instance, first-come first-served) the arbiter must include a mechanism for storing the order of arrival of the *seize* signals.

Figure 8.2 shows only the *seize* signals. In most systems a *clear* is also used. This permits a calling control unit to cancel a *seize* request before it has received the *accept* signal. It is also possible for an explicit *reject* (or *busy*) status signal, to be returned to the calling control unit.

184   CONTROL UNIT DESIGN

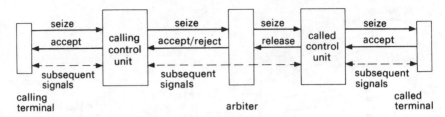

Figure 8.3   Signal exchange diagram for complete system.

The system described in the above example is not unique and many other arrangements are possible. The method described uses the concept of a single centre of control. When the control unit is *idle* the centre of control is the arbiter. Once the control unit is seized, the arbiter becomes subservient to the control unit, that is it may not return to the *idle* state until it receives the *release* signal from the main control unit. It is shown later that exceptions to the concept of centre of control are necessary in order to achieve a high system availability.

Figure 8.3 is the signal exchange diagram for a complete connection. It shows that, apart from the arbiter function, each part of the complete connection need only take account of terminals and control units associated with that connection. This separation of per-call functions and complete system functions is an important design concept which is especially relevant to computer controlled systems. It simplifies the design of the individual control units and concentrates in one area (in the arbiter) that part of the control which takes account of the rest of the system. It also simplifies modification of the control because the functions of the arbiter are usually invariant whereas changes in the control unit design are required during the life of a switching system as new services are added.

8.3   State transition diagram for a called control unit

We are now in a position to develop the state transition diagram for the called control unit. The first step is to decide on the precise signals that are to be passed between the control unit and interworking units. One suitable set of signals is shown in Figure 8.4a. The arbiter unit communicates with the control unit by *seize* and *release*, signals. Once the arbiter has seized the control unit, the subsequent signals that are sent to and from the calling control unit are *answer* and *clear* (in both directions). The signals passed between the control unit and the called terminal are *seize* and *accept*, together with *clear* in each direction. Note that a calling control unit receives *busy* from the arbiter and this is not apparent to the called control unit.

The process of developing the s.t.d. starts with the *idle* state, as shown in Figure 8.4b. At first sight it appears that the only valid event in this state (for a

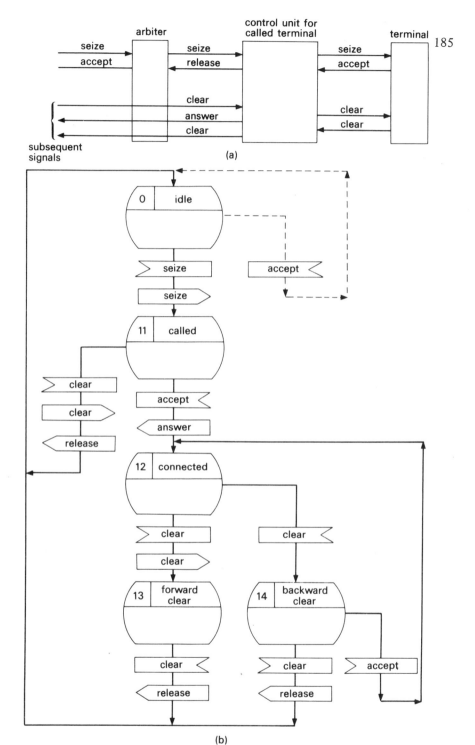

Figure 8.4 Design of called control unit. (a) Signal exchange diagram. (b) State transition diagram.

called only control unit) is a *seize* signal from the arbiter. The control unit responds by transmitting a *seize* to its terminal and moves into the *called* state. In the *called* state, the next event is normally the arrival of *accept* from the terminal; in this case the called control unit moves into the *connected* state and transmits an *answer* signal to the calling control unit. Another event that may be received in the *called* state is a *clear* from the calling control unit, indicating that the call is abandoned before it is answered. In this case the called control unit transmits a *clear* to the terminal in order to cancel the *seize*, and returns the arbiter to its free state by sending to it a *release* signal. The state of the called unit state then returns to *idle*.

In some systems it is physically possible for the *clear* signal sent to the terminal, and the *accept* sent from the terminal to 'cross in the post'. This implies that an *accept* could arrive after the called control unit has returned to the *idle* state. This possibility can be allowed for on the s.t.d. by adding the *accept* event with a path back to the *idle* state as shown in Figure 8.4b. Alternatively the s.t.d. may be drawn with the assumption that any event which occurs at a state for which it is not specified will be ignored. In practice the control unit is designed such that only the *seize* event is recognised in the *idle* state.

In the *connected* state, a *clear* may be received from either the calling control unit or the terminal. In the case of a *clear* from the calling unit, the called control unit transmits a *clear* to the terminal and waits for a *clear* back from the terminal before it transmits a *release* to the arbiter. Figure 8.4b is drawn on the assumption that the connection is under the control of the calling party and therefore, in the *connected* state, a *clear* from the terminal has no effect on the connection. A separate *backward clear* state is required as shown in Figure 8.4b. From this state an *accept* cancels the *clear* from the terminal and a *clear* from the calling unit permits an immediate release of the arbiter.

Both-way control unit. In many switching systems, such as a telephone system, a terminal may act either as a calling or a called device. In this case the control unit must be both-way. A *seize* signal from a terminal, when the control unit is in the *idle* state, indicates the calling mode. The control unit must then be busied to prevent an incoming call arriving at the terminal. Figure 8.5a shows how this can be achieved. The arbiter is provided with an additional input from its own control unit so that it may seize itself. This input competes with the other inputs for the seizure of the unit and there is therefore the possibility of simultaneous *seize* signals being received from other calling units as well as from itself. If the arbiter decides in favour of the other control unit, the control unit will be forced into the called mode.

The signal exchange diagram can be obtained by combining Figures 1.10 and 8.4a as shown in Figure 8.5b. This new diagram shows two simplifications.

(a)   There is no separate *accept* signal from the terminal and its control unit. Instead the *seize* signal is used for this purpose. That is transmission of a

STATE TRANSITION DIAGRAM FOR A CALLED CONTROL UNIT

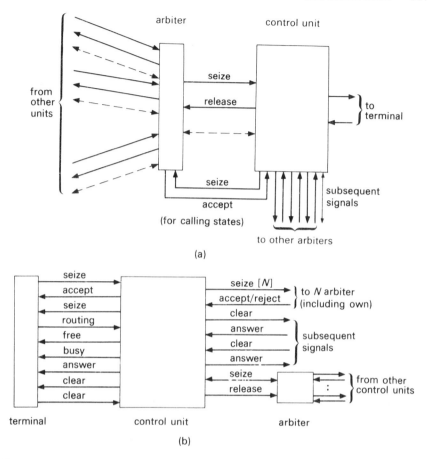

Figure 8.5 (a) Both-way control unit. (b) Signal exchange diagram for both-way control unit.

seize signal after the terminal has received a *seize* is regarded as an *accept*. (A practical example of this is a telephone system where the same electrical signal of placing a loop across the line is used to signal either *seize* or *accept* depending whether the terminal's control unit is in the *idle* or *called* state.)

(b) A single *seize* [N] signal is shown rather than N individual signals. The *accept* or *reject* signal is returned from the selected arbiter. The subsequent signals are then *clear* and *answer* in either direction.

The s.t.d. for the both-way control unit has more than ten states; this is awkward to present clearly on one sheet of paper. It is therefore desirable to adopt what some authors call a 'top-down' approach. The first step is to identify

188 CONTROL UNIT DESIGN

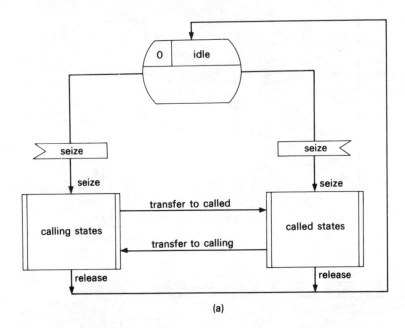

(a)

the major states of the control system and the relationships between them. For this control unit there are three major states: *idle*, *calling* and *called*, as shown in Figure 8.6a. This diagram also shows the possibility of a direct transition between *calling* and *called* states. The *calling* state to *called* state transition occurs if there is a simultaneous seizure from the terminal and another control unit. The *called* to *calling* transition occurs if, as a design objective, the control system moves directly into the calling mode when it receives a *clear* from a calling control unit. (In a telephone system this implies that dial tone is returned when an incoming call is cleared down under the control of the calling party.)

Figures 8.6b and (c) show example s.t.d.s for the set of calling and called states. As before, many variations in the details of these types of s.t.d.s are found in practical systems. In particular, Figure 8.6c shows the possibility of the direct transition from *called* to *calling* state. In this case there is no need for a *clear* signal to be sent to the terminal.

Reduced control systems. Before designing a control unit from an s.t.d. it is sometimes desirable to reduce the number of states. In the s.t.d.s discussed in this chapter there are a number of redundancies. For instance in Figure 8.6b the transmission of the *free* and *answer* signals to the terminal has no actual effect upon the calling control unit. The signals are merely relayed to the terminal. So, these signals could just as well pass through the control unit without it changing its state. This is shown diagramatically in Figure 8.7 where these signals are transmitted directly from the called control unit to the calling terminal. States 5

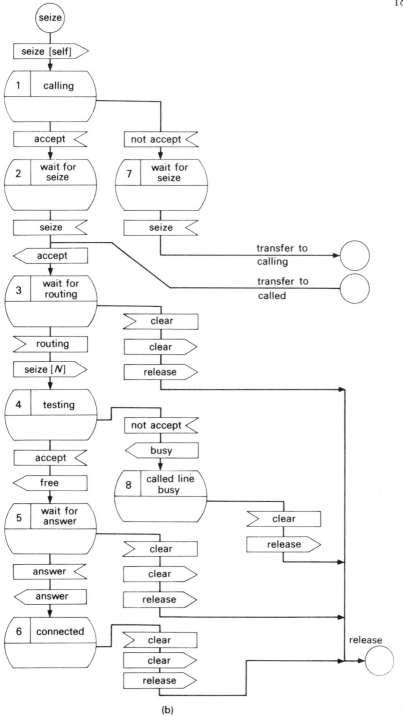

(b)

190 CONTROL UNIT DESIGN

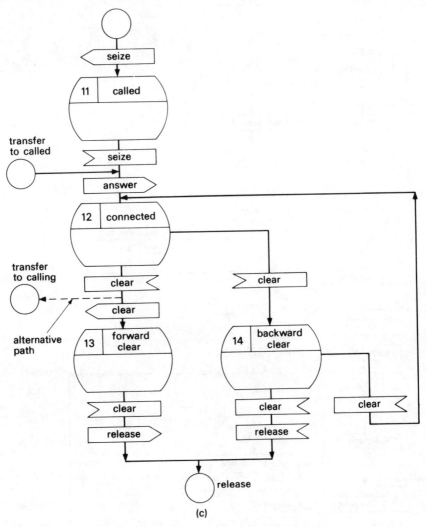

Figure 8.6 (a), (b) and (c) Design of both-way control unit.

and 6 are then redundant, and the only relevant events occurring in state 4 are *reject*, if called terminal is busy, and *clear* otherwise.

At first sight, state 14 in Figure 8.6c might appear redundant because the *clear* from the called terminal has no effect. This suggests that the *clear* and *seize* events from the terminal could be made to lead back into state 12. However, the reason that state 14 is introduced is to determine when the *release* signal is to be sent to the arbiter. When a *clear* is received from the calling terminal, a *release* can be sent immediately if the called terminal has already cleared. Otherwise the

STATE TRANSITION DIAGRAM FOR A CALLED CONTROL UNIT 191

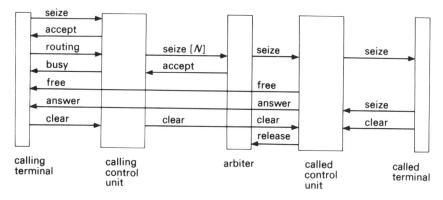

Figure 8.7 Signal exchange diagram for simplified control units.

control unit must wait for the *clear* signal from the called control unit. The function of state 14 is to record whether or not a *backward clear* had occurred. This state may be replaced by a test, which, when the *backward clear* arrives, checks whether the called terminal is clear. If it is clear, the *release* signal can be sent.

A similar modification removes states 2 and 7 of Figure 8.6b. Here the states are required to determine whether the arbiter has chosen a calling or called mode, in the case of simultaneous seizure by the terminal and another control unit. This information is not required until the *seize* from the arbiter has been received. The question can then be resolved by testing whether the arbiter has accepted the self-seize.

These modifications of the s.t.d. are all shown in Figure 8.8, where the total number of states in this example has been reduced to seven. It should be noted that the diagram is now more complex and possibly more difficult to understand. It is certainly less flexible because any future modifications to the functional requirements will be more difficult to achieve than with the simpler s.t.d.

Philosophy of state transition diagrams. The basic philosophy of a state transition diagram is that it is event driven. That is it assumes that a control unit is in a particular state and changes state only when an event occurs. A mathematical definition of the type of state we are considering is 'the class of all equivalent histories'. In other words, the state summarises all that a control unit needs to know about the previous history of events.

The use of s.t.d.s as a tool in the design of a control unit implies that the unit is designed to wait in each state until particular events cause it to move to another state. We have seen that there are occasions when an event occurs which has no immediate effect, but may modify the actions in later states. These may be regarded as events waiting for a state to occur. To keep to the philosophy of

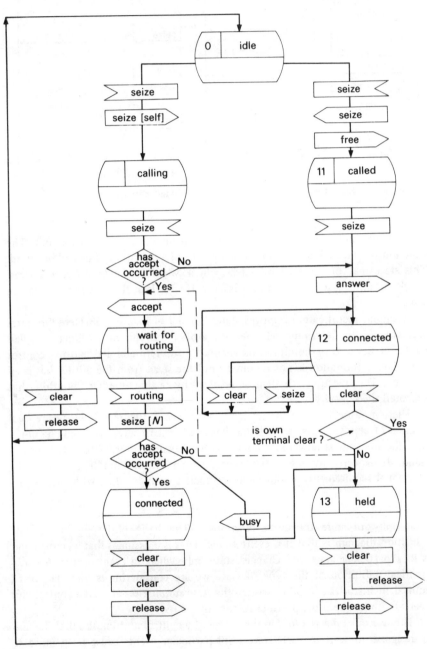

Figure 8.8 Reduced s.t.d. of both-way control unit.

SIGNALLING BETWEEN CONTROL UNITS   193

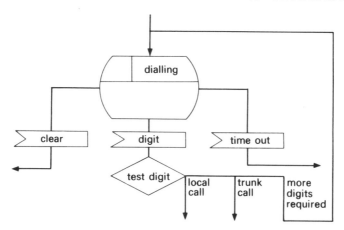

Figure 8.9   Detail of dialling state.

s.t.d.s such events have to be described as starting off a parallel set of states. One simple example is the *clear* from the called terminal shown in Figure 8.6c.

The number of states in this s.t.d. can be reduced if the events waiting for a state have the effect of setting a memory element which can be tested at a later state. This was the approach adopted in Figure 8.8. These memory elements must now be included in the total description of the state since they represent part of the relevant history.

So, in a practical application of s.t.d.s for design purposes, the states shown in the diagram are 'major' states and there is additional stored information available defining the state in greater detail. This stored information can be used for decision making. Another example is shown in Figure 8.9 where the major state is *dialling*, but the actual digits are stored and each time a new digit is received, a test is performed to see which the next state should be.

8.4   Signalling between control units

One of the results of resource sharing is that a control system is partitioned into a number of control units which are connected as required for a particular call. This gives rise to the need for a set of signals to be passed between the control units in addition to those passed between the terminals. A typical situation is shown in Figure 8.10 where a number of control units of (say) type A are connected to a smaller number of (say) type B when required (e.g. line units and main control units). In order to maintain a high system availability it is necessary to protect the overall system against two things:

(a)   *Misoperation of the system by users:* if a terminal seizes a resource but does not release it within some acceptable time, the overall grade of service is reduced. This occurs because the statistical assumptions upon which it

194   CONTROL UNIT DESIGN

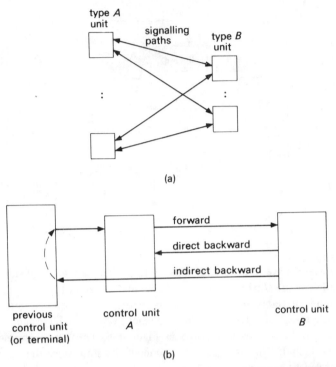

Figure 8.10   Interconnection of control systems: (a) Typical interconnection of control units. (b) Types of signalling path.

was designed are invalidated. It is desirable to provide a mechanism whereby a seized resource may be forcibly released from a seizing control unit if the resource is held for too long a time or receives an invalid sequence of signals. A force release produced as a result of an abnormally long holding time is referred to as a *time-out*.

(b) *Control unit malfunction:* in Figure 8.10a a faulty type A control unit may seize and not release a type B unit with the same effect as (a) above. On the other hand if there is a malfunction in a type B unit this prevents a connection being made for a proportion of calls (assuming random selection of type B units). However, if a type A unit can detect malfunctions in type B units it is possible for the type A unit to make an automatic second attempt, via an alternative type B unit. Then the system availability as seen by the user is increased.

So a desirable feature of a control unit is that it should be able to detect malfunctions in interworking units or misoperation by users. It must then be able to disconnect itself from the offending unit or terminal. It is therefore necessary to provide signals from which correct operation may be inferred and signals to effect a disconnection when required.

Signalling paths. Figure 8.10b shows three types of signalling path. In some cases only a forward path is provided. That is a control unit earlier in a chain of units may send signals to one later in the chain, but is not able to receive direct backward signals. An example of this can be seen in Figure 8.7. During the initial phase of a call the signalling between the terminal and the calling control unit is two-way, as they pass signals to one another. The terminal (or its user) can detect the absence of an *accept* (or reception of *busy*) and may clear. However, once the calling control unit has been connected to the called control unit, there is only forward transmission of signals between these two units. The called unit cannot directly affect the calling unit. It may protect itself from the calling unit by sending a *release* to its own arbiter but (as shown on the diagram) this will not be detected by the calling control unit. There is, however, indirect feedback in this case, since the terminal can detect the absence of expected signals from the called control unit and transmit a *clear*.

The provision of direct backward signals allows one control unit to protect itself against another. For instance:

- *acknowledge* when called control unit has successfully established a connection;
- *answer* direct to calling unit rather than indirect;
- *force release* to indicate that because of a malfunction or expiry of a time-out, the called unit wishes to disconnect itself.

When the first two signals are provided, a control unit always expects a signal and can therefore detect the absence of any signal as a malfunction.

In more complex cases, a whole range of acknowledge signals can be implemented so that a control unit early in the chain can more effectively detect a malfunction later in the chain (for example see the CCITT R2 signalling system described in Chapter 6).

8.5 Signalling techniques

Signals are generally transmitted between control units by the transmitter modifying one or more of a number of electrical conditions whose change is detected by the receiver. The conditions include logic, voltage levels, the amplitude or frequency of a tone and d.c. resistance between a pair of wires. There are two classes of signalling system, *time-independent* and *time-dependent*. In the time-independent class it is only the changes of the signalling conditions that have any significance. In time-dependent systems, the length of time between the changes is also significant.

The simplest type of signalling system is a time-independent system with binary-valued condition. Here the condition can be turned on or off, so the alphabet of signals is two:

signal name	condition
S	condition off to on
$\bar{S}$	condition on to off

However, the effective alphabet is only one since the signals must alternate, so at any one time there is only one signal that may be sent. (If the condition is already on all that can be transmitted is an off.) The variable is the time at which the signal can be sent. The simplest example is the *seize* and *clear* signals from one unit requesting service from another.

In general, a two-way signalling path is required. Figure 8.11a shows how a unit A can call for service from a unit B. There is a direct single binary-valued signalling condition in each direction.

The s.t.d.s for the two control units are shown in Figure 8.11b where it may be seen that a backward *clear* is used. A response is expected to each signal and this is known as a *handshaking* arrangement. The system availability can be improved by the addition of simple time-out checks. If an expected response does not arrive within some predetermined time, a malfunction is assumed to have occurred and suitable action is taken. Suitable action for unit A could be an automatic second attempt to seize another type B unit. The response of unit B to a time-out would be a *backward clear* signal.

This simple system works because the normal flow of control is from A to B only. B will never seize A although B can force release A once it is connected. In many cases, fully symmetrical operation is required. For instance a trunk circuit between two switching centres can be used for traffic originating from either end. This raises the problem of what happens when there are effectively simultaneous seizures from both ends. This is referred to as the *call collision* problem. A signal alphabet of only one signal is insufficient to resolve the collision problem. This is because after one control unit has transmitted a *seize* it is impossible for that unit to distinguish whether the next received signal is an *accept* of the transmitted seize or a *seize* from the other unit. Call collision is a particular problem with circuits which introduce significant transmission time delays such as transatlantic telephone cables with one-way delays of up to 50 ms or communication satellite circuits with one-way delays of around 300 ms.

The signalling alphabet can be increased by the provision of more signalling conditions or by the addition of time-dependent signalling. A second signalling condition provides a total alphabet of six signals, but in any one state there are only three alternatives that can be transmitted. These three alternatives are to change one or other of the two conditions, or to change both.

We will now show how an increased alphabet can be used for both-way use of a trunk circuit.

SIGNALLING TECHNIQUES 197

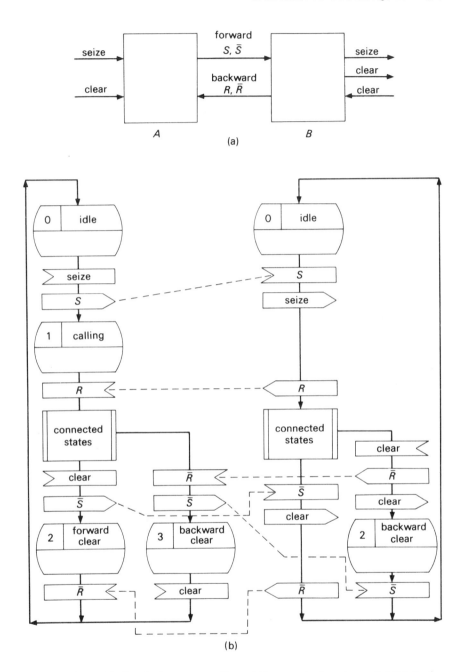

Figure 8.11 (a) Simple interconnection of time control systems. (b) S.t.d.s for simple compelled signalling system.

Time-dependent signalling. The alphabet produced with a single signalling condition can be increased from two signals by distinguishing between different durations of particular signalling conditions. For instance, an alphabet of four signals can be produced on an analogue circuit by a single sinusoidal tone with the following conditions:

Signal	Condition
A	Tone on for less than 200 ms
B	Tone on longer than 200 ms
C	Tone off for less than 200 ms
D	Tone off for longer than 200 ms

Further extensions to the alphabet are possible by increasing the number of distinguishable durations. More complex signals can be constructed from groups of simple signals. The most obvious examples are the dial pulses discussed in Chapter 1.

Persistence testing. In a practical signalling system it takes the receiver a finite time to recognise that a change of condition has occurred (A relay must have time to operate, a filter takes time to settle and so on.) In addition, noise on the signalling path or in the receiver may cause spurious changes. One very common form of noise in an electro-mechanical system is relay bounce. When a conventional relay operates, the contacts move together and hit one another. When this happens there is normally some mechanical reaction and the contacts briefly bounce apart, then come together again. This generally happens several times and the bouncing (rapid makes and breaks) generally lasts several milliseconds on a large relay (but much less on reed relays.) Bounce is also observed on release of relays, where it is caused by changes in the magnetic circuit as the armature moves.

In order to counteract the effect of noise (and bounce) in a signalling channel, it is necessary to establish that a condition change persists for a certain minimum period of time before it is recognised. For a completely electromechanical controller this is no problem because the inertia of the controlled devices smooths out the rapid changes or the noise burst. With electronic controllers it is necessary to provide such inertia explicitly. The persistence test may be performed as a part of the input function or in simple cases it may be built into the s.t.d. of the main control unit.

8.6 Design of a control unit for two-way traffic

As a design example, consider the specific system shown in Figure 8.12a. This system consists of two similar trunk control units (unit TCA and unit TCB) connected to either end of a trunk circuit which is used for traffic originating in

DESIGN OF A CONTROL UNIT FOR TWO-WAY TRAFFIC

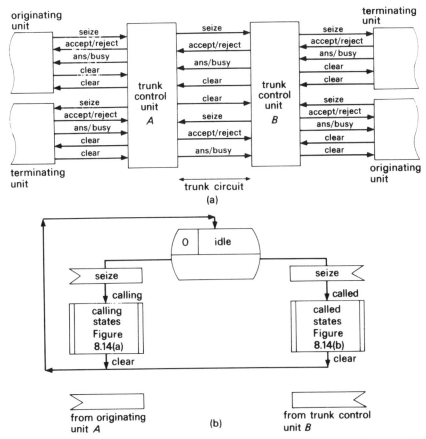

Figure 8.12 Control unit for control of both-way traffic on a trunk circuit. (a) Signal exchange diagram. (b) Top level s.t.d.

either direction. Each trunk control unit is connected to an originating and to a terminating control unit.

The specification is:

(a) Each originating unit may transmit a *seize* to its associated trunk control unit, and expects an *accept* or *reject* in response.
(b) The routing signal mechanism is not shown and is independent of this signalling system. All that is received by the incoming trunk unit is an *answer* or *busy* signal. On receipt of the *busy* signal, the trunk control unit must release the trunk circuit.
(c) The trunk circuit must be released by a *clear* from either end.
(d) In the event of a double seizure, each trunk control unit must send a *reject* signal to its associated originating unit.
(e) The originating units respond with a *clear* after receiving a *reject*.

200 CONTROL UNIT DESIGN

Figure 8.12b shows an s.t.d. for the major states which satisfies the specification. A trunk control unit is placed in the *calling* or *called* set of states by a *seize* from either the associated originating control unit or from the other trunk control unit.

Calling states. The first step in the detailed design of an s.t.d. is to produce the main 'thread' of states for the most common sequence of events. Consider first the main thread for the calling states of unit TCA Figure 8.13. This thread is entered by receipt of a *seize* from the associated originating unit. The first action is to transmit a *seize* to unit TCB and TCA enters state 1, *seize request*. The expected event in this state is the arrival of an *accept* signal from TCB which causes the transmission of *accept* to the originating unit TCA.

The routing signal is not shown as it will normally be sent over a separate signalling system. However, the usual result is, some time later, the reception of an *answer* signal forwarded by TCB.

This *answer* signal is forwarded by TCA to its originating unit and TCA moves to state 3, *connected (calling)*. The end of the call is normally signalled by a *clear* from the originating unit; this signal is repeated to TCB; and TCA moves into state 4, *clear F* (F for forward). Some time later TCB is ready for a new call and transmits a *clear*. This *clear* is repeated to the originating unit associated with TCA. TCA moves back to state 0, *idle*, and the main thread is complete. A similar main thread can be obtained for the *called* states.

The next step in the design is to consider, for each of the main thread states, all other possible events and their consequences. This often has the effect of introducing further states and the process must be repeated for the new states. Figure 8.14a shows the result of this process applied to Figure 8.13, as described below.

State 1: Seize request

(a) TCB may not be able to seize a terminating unit. In this case a *reject* rather than an *accept* signal is returned. The problem specification (*e*) calls for this *reject* signal to act as a backward clear, so TCA must move into a new state, (state 5) (*clear B*) where it waits for an acknowledging *clear* from A. When this *clear* arrives it is repeated to TCB and the originating unit TCA moves back to *idle*.

(b) It is possible for a *clear* to arrive from the originating unit before the *accept* (or *reject*) signal arrives from TCB. This indicates that the call has been abandoned. In this situation the *clear* must be repeated to TCB and TCA moves to state 4 (*clear F*) to await the acknowledging *clear* from TCB.

(c) When a double seizure occurs, both trunk control units transmits a *seize* and both enter state 1. Therefore a *seize* is a possible event in state 1. Recognition by a trunk control unit of a *seize* in response to its own transmitted *seize* indicates call collision. In our example, the specification calls for each trunk control unit to transmit a *reject* to its associated

DESIGN OF A CONTROL UNIT FOR TWO-WAY TRAFFIC   201

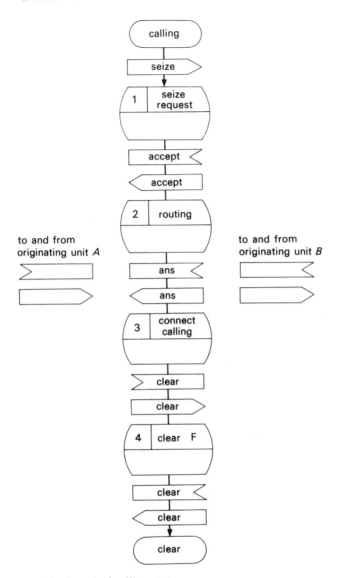

Figure 8.13 Main thread of calling states.

originating unit. Each trunk control unit moves to state 7 (*call collision*) until the expected *clear* arrives from the associated originating unit.

State 2: Routing
(a) A *clear* may be received from the originating unit if the call is abandoned before *answer* or *busy* is received.

202 CONTROL UNIT DESIGN

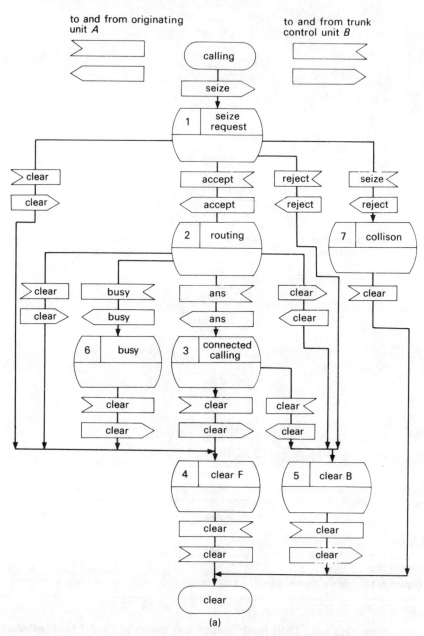

Figure 8.14 (a) S.t.d. for calling states for the trunk control unit A. (To provide protection against malfunctions, all states must be responsive to clear from other trunk control unit. All other events not shown are ignored.) (b) S.t.d. for called states for the trunk control unit B. (All states should respond to

DESIGN OF A CONTROL UNIT FOR TWO-WAY TRAFFIC 203

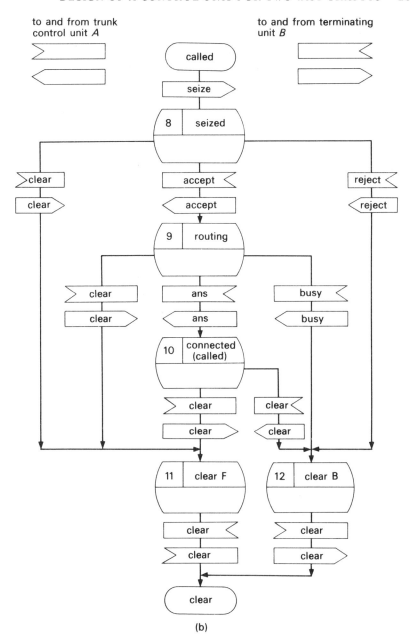

(b)

clear from control unit A. Also any source from originating unit B should respond with a reject but not change any of these states. All other events to be ignored.)

(b) *Busy* may be received from TCB which must have the effect of moving TCA to state 6, (*busy*). In this state TCA waits for a *clear* from its originating unit. This *clear* is repeated to TCB and TCA moves to state 4. (In a practical system, the *clear* from the originating unit will be sent as soon as the *busy* has been recognised. Thus the trunk circuit is released as soon as practicable. The busy indication to the calling terminal will be returned from the originating unit until a *clear* is received from the calling terminal.)

(c) the *clear* from TCB is dealt with below.

State 3: Connected (calling)

(a) A connection may be force released by a *clear* from TCB. This *clear* is repeated by the originating unit associated with TCA, and TCA moves to state 4.

Extra events. It is possible for a control unit in the signalling system to malfunction. This could have the effect of moving a trunk control unit into a state which was not considered in the design. It was mentioned in Chapter 2 that one protection against such malfunctions is to arrange that a control unit should always respond to a *clear* signal, even if that *clear* should not normally occur in that state. This precaution is achieved by adding the event *clear* from TCB to all states not otherwise having it. Figure 8.14a shows this extra event only for state 3. It is also required for states 1, 4, 5, 6 and 7.

At any time the trunk control unit responds to a *clear* from its originating unit immediately and moves to a new state. Because of time delays, TCB may have already transmitted a signal which is relevant to the old state of TCA but not the new. For example, a *clear* from the originating unit to TCA in state 1 moves TCA to state 4. If the *accept, reject* or *seize* have already been transmitted by TCB, they will arrive when TCA is in state 4. These 'late arrivals' may be taken into account by adding these events leading to the same states. Alternatively the control unit may be designed to ignore any specified events.

Figure 8.14b shows the detailed s.t.d. for the called states.

9
Some circuit techniques

9.1 Basic Elements

The relay. The basic elements required in a switching centre are:

— switches;
— a means of receiving signals from terminals and other switching centres;
— a control system, to operate the switches.

The switch elements are covered in Chapter 2. In this chapter practical examples of the elements of a control system and the ways in which they are used will be described.

A control system requires elements to:

— perform logical operations;
— store information;
— provide an interface between the control and the switching and information elements.

The signalling information may be sent in many forms, so a range of devices is required for transmitting and receiving the analogue quantities used in signalling and for interfacing them to the control system.

One component that can provide all the above functions is the relay, already introduced in Chapter 2. This may seem prosaic to many of the readers of this book who have been brought up on modern electronic technology. But it must be remembered that the vast majority (more than 95%) of the telephone terminals throughout the world are still controlled by systems based on such devices. Relays still have their place in the most modern electronic switching systems, so much of this chapter is devoted to studying their use. The remainder of the chapter discusses electronic techniques used to provide the same functions.

206 SOME CIRCUIT TECHNIQUES

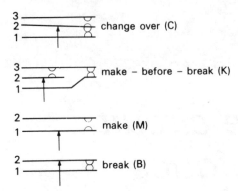

Figure 9.1 Contact arrangements of a relay.

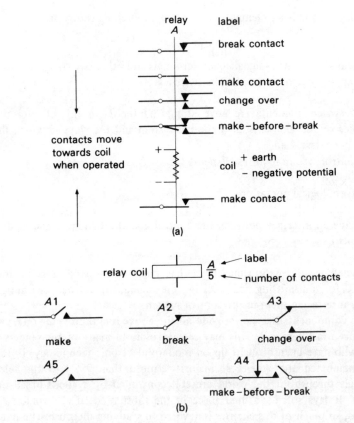

Figure 9.2 Circuit diagram conventions for relays. (a) Attached contact symbology. (b) Detached contact symbology.

BASIC ELEMENTS 207

The basic features of an ordinary relay were shown in Figure 2.10 (page 40). When the coil is energised, the magnetic field produced by the current through the coil attracts the armature and the movement of this armature operates the contacts. There are a number of different contact arrangements that can be used; some examples are illustrated in Figure 9.1. The operating mechanism for the contacts is made of insulating material and therefore all the contact sets are electrically isolated from each other and from the coil. Information about the detailed design of relays may be found in the references [1].

There are two main ways of representing a relay in a circuit diagram:

Attached contact symbology. Figure 9.2a shows a method whereby all the contacts of a particular relay are associated with the relay coil (or coils). The contacts are shown on the diagram with their direction of movement towards the coil; the action when the call is energised can be imagined by visualising the movable contact being pulled towards the coil. The negative supply is shown by − and the earth potential by +.

Detached contact symbology. The alternative method is that of detached contacts where the coil is shown as a rectangle together with a label (for example, *A*) (Fig. 9.2b). Under this label is a number indicating the number of contact sets operated by this relay. The contacts themselves are normally shown in the non-operated position and they are labelled to show which coil they belong to. If there is more than one winding in the coil, the rectangle is divided into the appropriate number of segments.

The use of attached contact diagrams is usually of value in the more complicated common control relay sets, but for simpler systems, and for circuit explanations, the detached contact method has its advantages.

It should be noted that most relay circuits use a negative polarity supply with respect to earth, typically -50 V. The positive side of the supply is earthed rather than the negative as this reduces the electrolytic action between the external cables and the ground.

Relays as interface elements. Relays make a convenient interface element as they can be operated by various signalling conditions. Their contacts provide on/off connections which may be used by the logic elements of the control systems. A simple example of a relay used as an input interface is shown in Figure 9.3a. The relay detects the loop condition used as a *seize* signal.

To receive dialled information a two-winding relay is needed, as shown in Figure 9.3b. Two windings are required because the transmission line to the switching system must be kept balanced. To maintain this balance there must be equal impedances from each wire to earth. So the relay acts as part of the transmission bridge as well as detecting the dial breaks and *clear* signals. The inductance of this relay is made high to minimise the attenuation of the speech signals, but the d.c. resistance must be relatively low to provide a high feeding

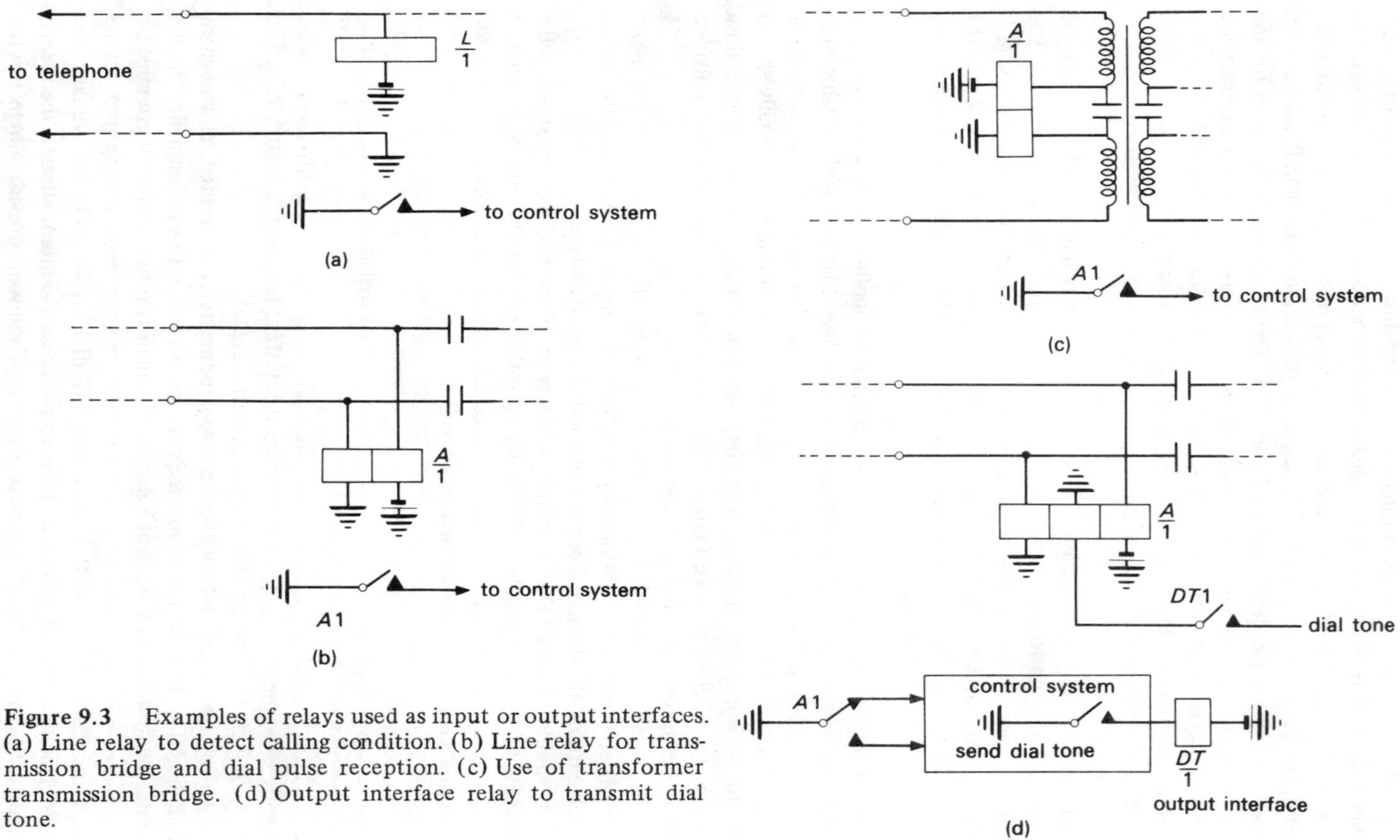

208 SOME CIRCUIT TECHNIQUES

Figure 9.3 Examples of relays used as input or output interfaces. (a) Line relay to detect calling condition. (b) Line relay for transmission bridge and dial pulse reception. (c) Use of transformer transmission bridge. (d) Output interface relay to transmit dial tone.

BASIC ELEMENTS 209

current of about 20 to 100 mA for powering the telephone instrument. Figure 9.3c shows a more expensive arrangement for a feeding bridge and interface which does not require the relay to have a high inductance. Standard references on telephony discuss the special design problems of this circuit for dial pulse reception [2]. These problems arise from the high resistance and distributed capacitance of a long cable.

Relays have direct application to output interfaces too. Audible signals such as busy tone may be transmitted to a telephone via a contact of a controlled relay. These audible signals must be balanced and a convenient technique is to add a third winding to the line relays. This winding then acts as a transformer as shown in Figure 9.3d. When a transformer feeding bridge is used, the tones can be transmitted via an extra winding on the transformer.

Relays as logic and memory elements. Relays can be used for combinational circuits. In fact, much of what is known today as 'switching theory' in the context of logic design, such as application of Boolean algebra, Karnaugh maps and so on, was developed originally by relay set designers [3]. The variety of contacts that can be used on a relay gives considerable flexibility to the logic manipulations that they can perform.

An electrically latching relay can be used as a memory element, equivalent to a 'flip-flop'. The latching is achieved by using one of the relay's own contacts to provide a holding path as shown in Figure 9.4a. There are several ways to release an operated relay:

— The hold path may be broken as in Figure 9.4a;
— The energising current may be shunted as in Figure 9.4b;
— Another winding may be provided which, when energised, cancels the magnetic field and thus causes release of the relay, as in Figure 9.4c.

As a example a 4-bit store is shown in Figure 9.5. When an earth is applied to the 'store' input, any pattern of earths on the four input lines operates the appropriate relays. Contacts on the storage relays provide earths for the operated relays. Once the 'store' input is released, the storage relays remain operated until an earth is applied to the 'reset' input.

Timing. Relays can be used to provide timing elements by connecting capacitors with values of the order of several tens of microfarads in parallel with their coil. Depending upon the contact arrangement, the relay can be made slow to release, or slow to operate and slow to release. With this method it is easier to obtain longer release times if very sensitive relays are used. Release times of up to 30 seconds are feasible with capacitors of 200 μF and coils of 10 000 Ω.

When shorter operate or release delays are required, they can be obtained by modifying the magnetic circuit of the relay. If a collar of copper (called a *slug*) is

210   SOME CIRCUIT TECHNIQUES

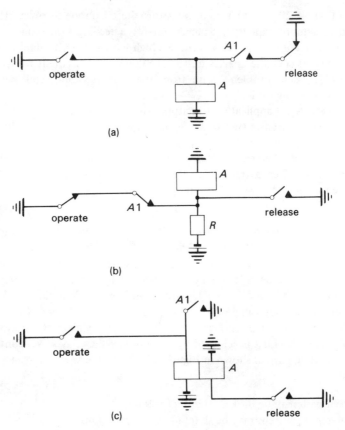

Figure 9.4   Use of a relay as a memory element. (a) Direct operation. (b) Shunt operation. (c) Double winding.

fitted around the coil of a relay, it acts as a single-turn short-circuit winding. When the current of the relay coil is turned on or off a high current is induced in this copper ring by the changing of the magnetic flux linked to the ring. This induced current produces a magnetic field which opposes the change causing it and therefore tends to maintain the field in its previous state. When the collar is placed at the armature end (a front-end slug) the relay is slow to operate and slow to release. If the collar is placed at the other end (a heel-end slug) the relay operates at the normal speed, but is slow to release. Operate lags of up to 100 ms and release lags of up to 600 ms can be obtained by this technique. Slow releasing can also be achieved with a low resistance winding on the coil which is short circuited by a contact on the relay.

If slow-to-operate times in excess of 30 seconds are required, they can be provided with a thermal relay in which the movable contact consists of a

BASIC ELEMENTS 211

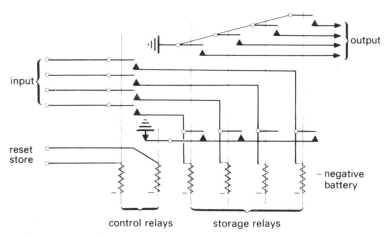

Figure 9.5 Use of relays for a 4-bit store.

bimetallic strip surrounded by a heating coil. Application of current to the heating coil causes the bimetallic strip to heat up slowly, and after a suitable delay, it makes the contact.

Translators. Most switching systems require large look-up tables in order to decode routing digits into suitable routings and to associate particular cable-pair terminations with particular dialled numbers. In an electro-mechanical system the most common form in which such tables are realised is the *distribution frame*. In a system block diagram this item may look insignificant but in practice these frames occupy a significant percentage of the total floor area required for a large switching centre. The principle of the device is simple. It generally consists of a two-sided rack. On one side the cables from the terminals are connected to terminal blocks. The other side is connected to the switching centre. The connections on this side are generally arranged in directory number (d.n.) order. Any external cable can be connected to any switching centre termination by means of jumper wires. The resulting mass of wiring is therefore a translation table. The advantages this provides are:

— a convenient termination point for all cables, whether in use or not;
— the ability to allocate directory numbers irrespective of cable-pair order and to rearrange cable routes without changing directory numbers;
— a convenient point to cross-connect permanent circuits which do not need switching;
— a convenient point to isolate the line from the centre for testing purposes and to provide a fuse (and a lightning protector if the line has a portion of its travel on an overhead route).

212 SOME CIRCUIT TECHNIQUES

In many electro-mechanical switching systems, especially step-by-step it is desirable to provide another degree of flexibility, so that the connections to the system need not be in d.n. order. This is normally required so that:

— the traffic can spread evenly over all the system inlets;
— lines requiring special services can be connected to special inlets if necessary;
— lines can be arranged if faults occur or additions are made to the switching equipment.

The inlets to the actual switching machine are numbered in some way. The numbers used for this purpose are called *equipment numbers* (e.n.). It is therefore necessary to provide a second frame to provide the d.n. or e.n. translation. This is usually of similar construction to the main distribution frame (m.d.f.) and is called the *intermediate distribution frame* (i.d.f.).

Similar (but smaller) frames are required for route translation. In all switching systems a main distribution frame is needed, if only to provide a place to terminate cables. However, in electronic and semi-electronic systems, most other translations can be performed more compactly. One type of translator that is still in extensive use is the Dimond ring store [4]. This is shown in Figure 9.6 applied to a d.n. or e.n. translator. The objective here is to translate a four digit dialled code into a number which tells the system the physical location of the line unit corresponding to the dialled code.

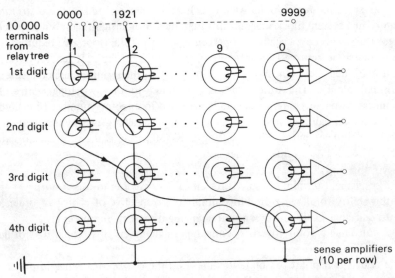

Figure 9.6 Dimond ring translator (sense amplifiers only shown on 10th column).

The translator consists of a number of magnetic rings of typically about 5 cm diameter. Each ring is provided with a coil connected to a sense amplifier. The cores are threaded with wires which act as a single turn primary. When a pulse of current is transmitted along one of the wires, an output is generated from the sense amplifier of each coil that it threads. In Figure 9.6 it is assumed that the equipment number is specified itself by a four digit number. There are therefore four rows, each of ten cores. There is one wire for each directory number and each wire threads the cores appropriate to its translation. The wires shown in Figure 9.6 would produce a translation of 1222 for an input of 0000 and a translation of 2120 for an input of 1921.

9.2 Electronic components

Electronic components are ideal for logic and memory devices. There is a vast range of suitable devices, ranging from simple logic gates to specialist sub-systems produced by large scale integration. It is practical to construct a push-button dial telephone which generates loop-disconnect signals using basically a single device. This device decodes and stores the sequence of button depressions and converts them to the correct time sequence of loop disconnections. The same device can also be programmed by the user to remember ten frequently used numbers. These can be recalled by use of only one or two depressions of the appropriate buttons. In relay technology these same functions would require about 150 relays. Today they fit into the standard telephone instrument.

Of even greater application are programmable devices. These provide the potential for flexibility as described in the previous chapter. The details of electronic devices will not be described here; they are amply covered in many texts.

Output to electro-mechanical devices. In an analogue telephone system it is at the interface between the electronics and the still necessary electro-mechanical portion where most of the technology problems lie.

The obvious interface between electronic and electro-mechanical sub-systems is the transistor as shown in Figure 9.7a. The transistor is controlled from a bistable which is turned on or off by the logic. A practical circuit is more complex than is shown in Figure 9.7a because protection circuits are required to prevent the transistor from being damaged by the back e.m.f. produced when the relay is turned off. An alternative to the transistor is an opto-electric isolator. This has the great advantage of providing complete electrical isolation between the electronic and electro-mechanical circuits.

The provision of a bistable and driver for each relay can be expensive. If there is a large number of relays, but only a few need to be operated or released in a given time, it is possible to use bistable relays operated and released by the outputs from a relay decoding tree or from an electronic matrix decoder. For

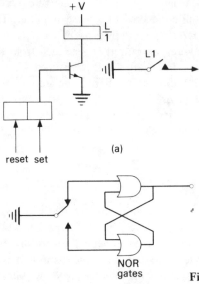

Figure 9.7 Interface between electronic and electro-mechanical elements.

example, 1024 relays could be controlled by a set of relays with only 10 electronic inputs (plus one to specify operate or release). In many systems this is more economic than providing 1024 electronic interfaces. The disadvantage is that it takes a significant time to operate or release one relay (say 10–100 ms, depending upon the type of controlled relay) so there is a limit to the number that can be controlled. Also the timing of individual relays may not be accurate because their operation or release may have to wait until the driver has served other requests. This type of system is known as a *slow driver;* the type with individually addressed bistables is known as a *fast driver*.

Input to electronics. In the reverse direction, the state of a relay contact can be converted to a logic signal by the circuit shown in Figure 9.7b. This is what is known as an *R-S* flip-flop. When the contact moves across it changes the state of the flip-flop. One of the characteristics of relays is that their contacts bounce on operation, and (more surprisingly) on release. This bounce can last for several milliseconds with some of the larger relays. In the circuit shown, the bounce is removed if the movable contact only bounces against its new position and does not oscillate between the fixed contacts. If this should occur it is necessary to provide timing circuits to ignore short transitions. Alternatively a logic circuit can be used to cause such transitions to be ignored. It is possible to obtain bounce-free operation with mercury-wetted contacts, but these are generally uneconomic for all but very specialist applications.

An alternative device to a relay for the input of line conditions is the ferrod [5]. This is a saturable ferrite-core transformer. The line current feeding a telephone line flows through two windings on this transformer and causes it to saturate. Each ferrod has a sense winding and a drive wire. When a pulse of current is passed through the drive wire it produces an output in the sense winding. However, if the line current is flowing, the transformer becomes saturated and no output is then produced. Thus the device can be used as a simple loop detector.

9.3 Realisation of arbiters

Scanner. Any switching system must contain at least one arbiter since there will always be some operations that must be performed 'one-at-a-time'. Arbiters in electronic systems are straightforward to construct. The simplest type is a *scanner* as shown in Figure 9.8. This shows a general arrangement where several sources require access to a limited number of servers. Since the sources can make requests for service at any time, an arbiter is required to:

— avoid two sources seizing the same server;
— (in this case) permit a one-at-a-time common control to operate switch network.

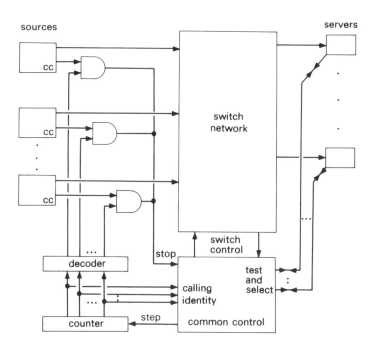

Figure 9.8 Principle of scanner controlled arbiter. (cc represents calling condition from source. This is assumed to be removed once switch has connected source to a server.)

216 SOME CIRCUIT TECHNIQUES

Both functions are served by a single arbiter in Figure 9.8. Each source provides an output indicating a calling condition and each output can be connected in turn to the common control by a decoder driven by a counter. Each time the control issues a *step* instruction, the counter steps on and addresses the next source. At the end of a scan, the counter returns to the first source.

This scanning proceeds until a new calling condition is detected and a *stop* signal is received by the control. The address in the counter then identifies the calling source.

Each server provides the control with a test lead which has a potential indicating when the server is available. The control selects one of the free servers and instructs the switch network to interconnect the calling source and selected server. This server then removes the potential indicating it as free. The connection of the switch path removes the calling condition and the scanner can restart and look for calling conditions. Thus if two or more sources call simultaneously, only one will be processed at a time. (If no server is available, in this simple system, the control waits until one does become free.)

Thus any control system which is driven by a clock and 'looks' for work automatically provides the arbiter function because it can recognise only one event at a time, even though several events may have occurred simultaneously.

Parallel arbiter.   In electro-mechanical systems and more advanced electronic systems parallel arbiters are required. Electro-mechanical systems cannot use scanning because the mechanical wear (and acoustic noise) would be unacceptable. Electronic distributed control techniques also require parallel rather than serial arbiters as these enhance the overall system reliability.

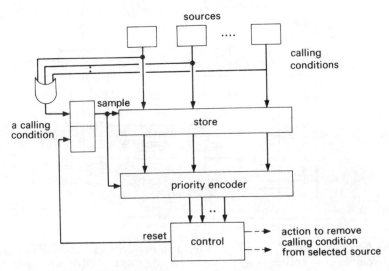

Figure 9.9   Principle of parallel arbiter.

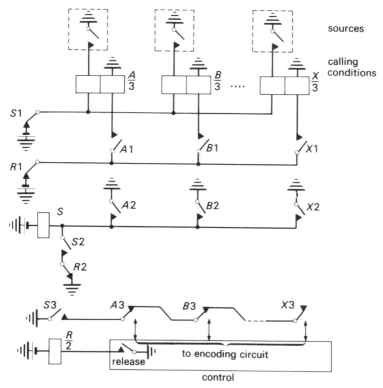

Figure 9.10 Simplified diagram of parallel arbiter constructed from relays. The principle of operation is that one (or more) of the sources requiring service places an earth on its calling wire. This operates the corresponding relay A, $B \ldots X$. The second contact on the relay operates the group relay S which remains held by contacts $S2$ and $R2$. Operation of relay S disconnects all the calling wires from the relays so that any subsequent calling condition will have no effect upon the relays $A, B, \ldots X$. The relays which are operated remain operated via one of their own contacts and the second winding. In this example a simple priority order is achieved using the third set of contacts. If several relays are operated only the left-most one provides an input into the encoding circuit. The system is then stable. When the control has completed its actions, relay R is operated briefly to release all the held relays. A practical circuit is slightly more complex as care is required on the timings of the various operations discussed.

The general technique of a parallel arbiter is shown in Figure 9.9. Each source, as before, provides an output to indicate a calling condition. These outputs are ORed together. If a calling condition occurs, the states of all the conditions are sampled and stored. A circuit (the priority encoder) takes these outputs and produces, in coded form, the address of the unit which is calling. If there happens to be more than one calling condition, the circuit selects only one on some priority basis. The control then takes action to connect the calling

218 SOME CIRCUIT TECHNIQUES

source and thereby remove its calling condition. After the control has dealt with the selected source, the arbiter is reset. If further calls have arisen during the time taken by the control to perform these actions, the arbiter is immediately reactivated.

In computer terms the circuit is known as a priority encoder and the overall arbiter is known as a non-preemptive interrupt mechanism.

A relay circuit which performs the same functions is shown in Figure 9.10.

9.4 Third wire control

A common feature of all electro-mechanically based switching systems (and also many electronic ones) is the use of a third control wire switched in parallel with the two (or sometimes four) speech wires. Typical names given to the control wire include

P-wire (short for private-wire) in U.K. Strowger systems;
c-wire in some crossbar systems (the speech pair being called a and b).

The primary function of the control wire is to indicate whether or not a terminal or a resource is available. A secondary function is to hold the path through a switch network. These two functions are necessary for any electro-mechanical system. There are other functions which may be implemented via the control wire and some of these are discussed later with reference to specific systems.

Typical coding for the control wire is:

> Earth potential — busy
> -50 V   — available
> open circuit   — not available.

Other codings are in use. A common alternative is to use open circuit as available and to use earth potential for both busy and not available.

Line unit. Typical use of the control wire in a functional divided system is illustrated in Figure 9.11a. The line unit itself is shown in detail in Figure 9.11b. The sequence of operation is as follows:

(a) When the line unit is *idle*, the K and L relays are both unoperated. There is therefore a -50 V potential on the c-wire (via the K relay).

(b) A *seize* from the terminal is signalled by a loop across the line which operates the L relay. The $L1$ contact forwards the *seize* to the access switch control.

(c) The access switch control must immediately change the state of the line unit from *idle*. This prevents another control unit from connecting an

THIRD WIRE CONTROL 219

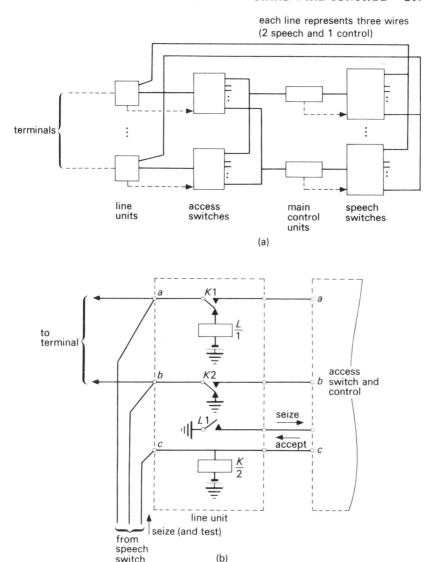

Figure 9.11 (a) Typical arrangement of control wires in functionally subdivided system. (b) Details of typical line unit.

incoming call to this terminal. In Figure 9.11b the change of state is produced by an *accept* signal from the access switch control. This signal is typically placing an earth on the *c*-wire, which has the following effects:
— indicating to any other control unit that this line unit is busy;

- operating the *K* relay and connecting the speech wires through to the access switch;
- removing the calling condition to the access switch (because the *L* relay is disconnected from the line).

(d) The line unit is then in the *connected* state and remains so until it receives a *clear* signal from the access switch control. The *clear* signal is the removal of the earth on the *c*-wire by the access switch.

(e) Each of the main control units has an associated *c*-wire to indicate its availability. On receipt of the *seize* signal from the line unit the access switch tests the *c*-wires and selects one that is available. It must immediately transmit a *seize* to the selected main control unit which may in turn change the *c*-wire potential to indicate not available. (Some circuit details of this operation are discussed below.) The main control unit then returns the *accept* signal to the calling terminal and awaits the *routing* signal.

(f) This *routing* signal is decoded and the speech switch is operated to make all three wires from the selected line unit available to the main control. The main control tests the potential of the *c*-wire. If it is -50 V, the called terminal is free and the control unit immediately places an earth on the *c*-wire of the selected line unit. This has two effects:
- the called line unit then indicates 'busy' to other main control units;
- the *K* relay operates so that when the called terminal answers there is not a calling condition into its own access switch because the *K* relay has disconnected the *L* relay.

(g) If the main control unit finds that the *c*-wire of the called line unit is at earth potential, the path through the speech switch is cleared and a *busy* signal is returned to the calling terminal from the main control unit.

(h) The centre of control for an established call is then in the main control unit. When the main control unit determines that the call is completed, it removes the earths from the calling and called line units to restore them both to the *idle* state.

Access switch control. In a function-divided system such as that shown in Figure 9.11a it is possible for a main control unit to receive simultaneous *seizes* from several access switches. When there are many sources (access switches) and servers (main control units) it is uneconomic to provide each server with a parallel input arbiter, such as that described in Section 9.3.

A more economic technique is for each server to broadcast its available/non-available status by an appropriate potential on its control wire. When an access switch control decides that a server is required, this control tests the status of the server control wires. When it finds a control wire with an available indication, the access switch control immediately transmits a *seize* to the selected server. On receipt of the *seize* the server changes its availability indicator. With this technique there are two problems:

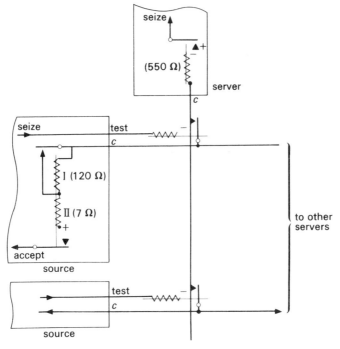

Figure 9.12 Typical realisation of simplified arbiter.

(a) the server has no direct information about the identity of the seizing source;
(b) the process of detecting an available condition, transmitting a *seize*, and subsequently removing the availability condition, takes a finite time. With relay circuits this time can be as much as several hundred milliseconds. It is possible for double connections to occur if, during this unguarded time, a second source tests the condition of the selected server and decides to transmit its own *seize*.

The probability of double connection can be reduced by decreasing the unguarded interval. A technique which does this involves providing an analogue circuit which detects the presence of two *seizes* or two tests. Figure 9.12 shows the principle used in Strowger step-by-step systems. This diagram shows two access switches connected to one server. The server is available if there is a negative potential on the control wire and unavailable if there is an earth. The access switches test the state of the server by operation of the appropriate relay. This relay connects a two-winding test relay to the control wire. If the server is free, the sensitivity of the relays is such that both the relays in the access switch and in the server operate. The relay in the access switch indicates to its control that the *seize* has been accepted and therefore a connection can be made. A second contact on the access switch test relay, short circuits the high resistance winding. The sensitivity of the relay is such that it remains operated but the potential on the control wire is then nearly zero, so any attempted *seize* by another access switch will be unsuccessful. The sensitivity of the test relays is chosen so that neither operates if two access switches try to seize the same server simultaneously.

Release of the connections may be made from either end. The server releases by disconnecting its test relay from the control wire. This releases the test relay in the access switch. The access switch releases by disconnecting its test relay. (A detailed analysis of this circuit may be found in Reference [6].)

9.5 Junctors

The main functions required in a telephone system junctor are to provide

(a) a speech path (and possibly a through signalling path);
(b) a feeding current for the telephone instrument,
(c) detection and insertion of signalling conditions; and
(d) access to other sub-systems.

The essential parts of a relay type local junctor are shown in Figure 9.13. The speech path is provided by a conventional capacitor transmission bridge. A bypass switch is included to divert the incoming pair to a register access switch for reception of digit signals. Busy tone, number unobtainable and ring tone are transmitted via the A relay. On the B side operation of the R relay connects ring current to the called line. The ring current passes through one winding of the C relay. This relay is designed to be slow to operate to prevent its operation by an a.c. signal. When the called terminal is answered a d.c. path is completed and the direct current operates the C relay. This action is known as *ring trip*. Operation of the C relay immediately removes the ring current (and the ring tone to the calling user) and the speech path is completed. Relay C is held operated by a second winding energised by contacts $R3$ and $C3$.

Almost every analogue-based switching machine requires a circuit similar to that described. Although some electronic replacements have been developed for d.c. feeding and for ring current supply and trip, none has yet met the stringent requirements for robustness of a circuit which is directly connected to telephone cables. The main danger is that cables can suffer very high induced voltages (sometimes exceeding a hundred volts) when a lightning strike occurs in the vicinity of a cable. For the older type of overhead wires on telegraph poles, there is the additional hazard of a direct lightning strike. Thus even the most modern (analogue) computer controlled switching machines still use relays such as shown in Figure 9.13 to complete the interface between the telephone line and the rest of the switching system.

JUNCTORS 223

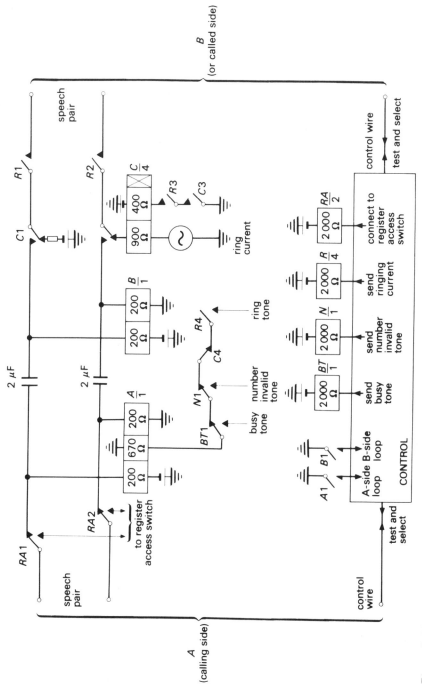

Figure 9.13 Local junctor constructed from relays (some typical resistance values shown).

224 SOME CIRCUIT TECHNIQUES

The control is not shown in detail as this varies considerably between systems and a wide range of technologies may be found. The older systems use special mechanical devices associated with the selector mechanism. Other technologies include various forms of relays, electronic devices and micro-processors.

Signalling receiver

Dial pulse receiver.   As an example of electronic circuit techniques, Figure 9.14 shows a practical circuit for decoding dial pulses into decimal numbers. This circuit can be used for individual receivers. Time-shared designs can be achieved by scanning a number of contacts associated with different registers at a rate fast enough not to miss the shortest change. Typical scanning periods are between 5 and 10 ms.

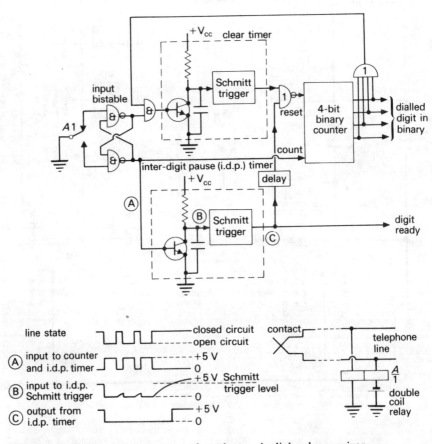

Figure 9.14   Simplified diagram of an electronic dial pulse receiver.

M.F. push-button signals. The digit signals from a multifrequency push-button telephone consist of pairs of tones, one selected from four in a low frequency band and one selected from three (or four) in a high band (see Ch. 2). The major problem in designing receivers for these signals is the possibility of signal imitation. While a subscriber is keying a number, the handset is off the hook. Between depressions of the buttons, the microphone is connected to the line, so speech and background noise or music enter the transmitter and can simulate a valid signal, thereby causing a misrepresentation of the digits at the switching centre. A typical objective is to require that misrouting due to this cause occurs for less than 1 in 10^4 of calls made. The signalling frequencies have been chosen to minimise this possibility, but two other techniques are used to minimise it further.

(a) Speech energy normally covers a wide spectrum of frequencies, so a guard technique can be used which inhibits detection of a signal if significant energy is present at a frequency other than two of the valid signalling frequencies. Figure 9.15 shows a typical receiver (developed originally by Bell Laboratories) [7] which achieves this requirement in a very elegant manner. There are filters for each of the signalling frequencies but each band is preceded by a band *elimination* filter and limiting circuit. The band elimination filter removes the frequency of the other band and hence, in the absence of noise or voice, the only signal left is a single sine wave. This is squared by the limiter and triggers one of the outputs if it falls within the bandwidth of the individual filters. The limiter produces an output with a maximum amplitude. Hence, if more than one frequency is present at the input to the limiter, the output energy is dispersed and therefore none of the filters will respond. This proves to be a very effective method of achieving high immunity to voice imitation. Note that a dial tone filter is also required in this receiver.

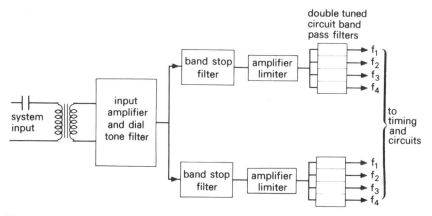

Figure 9.15 Bell system receiver for a multifrequency push-button telephone.

(b) The second technique is persistence checking. The logic at the output of the receiver includes timing circuits which give an output only if a valid combination of tones persist for longer than 33 to 50 ms. They also time the inter-digit pause for a minimum 33 ms. The maximum signalling rate is therefore between 10 and 15 digits/second.

Trunk (m.f.) signalling. A typical receiver consists of five or six filters connected to the line. The outputs from the filters are rectified and input to a comparison circuit which checks first that there are two and only two tones present, second that they have amplitudes within typically 6 dB of each other (and above a minimum) and third that a valid combination (and gap) persists for longer than 40 to 50 ms.

Voice imitation is not a problem for trunk signalling because the speech path is not connected to the receiver. However, trunk m.f. receivers are subject to voltage surges and noise bursts when their associated circuits are switched. In order to minimise the effect of false reception due to these causes, a signalling system normally requires the use of a prefix combination before a code, or a system of acknowledgements (as is described in Ch. 6).

9.6 Switch matrix control

An $M \times N$ switch matrix which is built from an array of electrically held relays contains MN relays. Direct operation of each relay is uneconomic and a co-ordinate operation technique is usually used. Figure 9.16 shows a typical form. Each of the MN relays has an operate winding connected to one of a set of horizontal operate wires and to one of a set of vertical operate wires. A diode is required in series with each operate winding to prevent the energisation of unwanted coils. An earth applied to a vertical wire and a negative potential applied to a horizontal wire operates only the relay connected to both wires. For an electrically held relay an additional winding is required to hold the relay once the operate conditions have been removed. This hold winding is connected via its own contact to the control wire. The relay remains operated as long as the control wire is held at earth potential. This arrangement has the advantage that the connection is released automatically when the earth is removed from the control wire.

Magnetic latching reeds. Magnetic latching relays can be built with 'square loop' magnetic materials. In principle, a particular relay can be operated or released by sending a pulse of current $½I$ along a vertical and horizontal where I is the current required to operate (or release) a contact. This method requires bipolar current generators. A simpler method, which uses only unipolar current generators is that used in the Bell Laboratories Remreed switch matrices in their No. 1 and No. 1A ESS switching systems [8].

Each relay is placed in a hole in a steel plate. This plate acts as a barrier to magnetic fields and allows each member of the reed contact pair to be magnet-

SWITCH MATRIX CONTROL 227

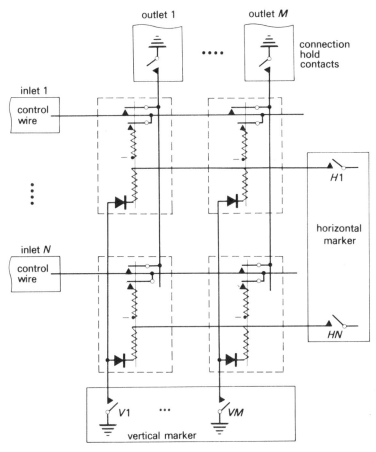

Figure 9.16 Co-ordinate operation of electrically held relay crosspoint matrix (control wire only shown).

ised separately. There are two coils on each relay. One coil has N turns on one side of the plate and $2N$ (in the opposite direction) on the other side of the plate. The other coil has $2N$ turns on one side and N (again in the opposite direction) on the other. The two windings are shown in Figure 9.17 and are called the vertical and horizontal windings by analogy with the co-ordinate selection technique. The fields produced when a pulse of current is passed through the winding are also shown.

A pulse of current in only one winding magnetises the two blades in opposite directions. Therefore the contacts release (or remain released). However, when a relay is energised by current in both coils, the larger field in each half predominates and the two blades are magnetised in the same direction. They then attract and close.

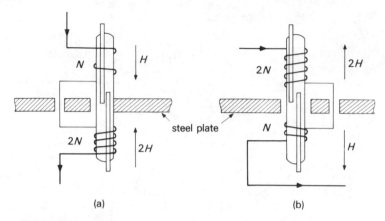

Figure 9.17 Windings on Bell System Remreed. For this device, $N = 31$ turns and a current of 4 A is required. The switching time is 4 ms. The devices are commonly arranged in 8 x 8 matrices and eight winding in series have typically 23 Ω resistance and require 96 V drive. (a) Vertical winding. (b) Horizontal winding.

This matrix therefore has the convenient property of self-clearing as the setting of a crosspoints automatically resets all other crosspoints on the same horizontal and vertical as the selected crosspoints.

Extension to multi-stage networks. Co-ordinate selection can be extended to any number of switching stages. Figure 9.18 shows as an example the organisation of a common control three-stage access switch. Only the control wires are shown. The common control can selectively access groups of these control wires by a relay known as a *connector* (Fig. 9.19). In Figure 9.18 it can be seen that the operation of a particular connector relay gives the common control access to all the control wires associated with the outlets of a particular A-stage switch or the inlets associated with a particular C-stage switch.

The principles of operation of this system are as follows:

(a) A line unit detects a calling condition and sends a seize signal to the *line identifier*.

(b) The *line identifier* determines the identity of the calling line (for future reference) and seizes the *control sequencer*. Once the *line identifier* has detected a calling condition, it ignores any further calling conditions until the original call has been connected.

(c) When the *control sequencer* has accepted the call from the *line identifier*, the *line identifier* sends a signal to the calling line unit to place a mark condition on its associated inlet.

SWITCH MATRIX CONTROL 229

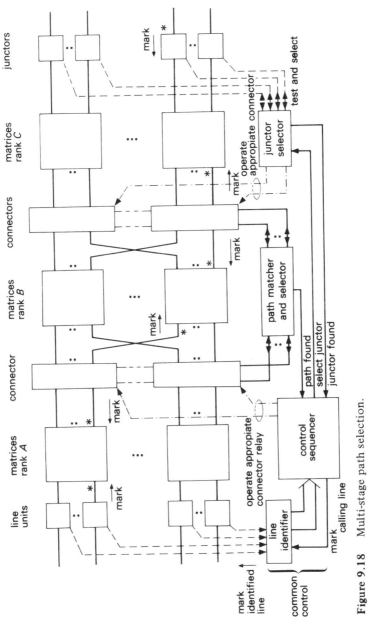

Figure 9.18 Multi-stage path selection.

230   SOME CIRCUIT TECHNIQUES

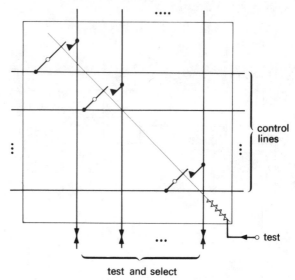

Figure 9.19 A connector relay.

(d) The *control sequencer* must select a free junctor. The *junctor selector* system has access to the availability indicators of all the junctors and, on command from the *control sequencer*, selects a suitable free junctor. A signal is sent to the selected junctor to mark the outlet of the switch to which it is connected.

(e) The next task is to find and operate a path between the calling line unit and the selected junctor. This requires a choice of a B-stage matrix. To effect the decision, the *control sequencer* operates the connector associated with the A-stage matrix used by the calling line unit. The *junctor selector* operates the connector associated with the inlets of the C-stage matrix serving the selected junctor. The two sets of outputs from the two operated connectors are input to the *path matcher and selector*. This sub-system checks the availability potentials on the control wires to discover which links are free and thereby determine which B-stage matrices are usable for the connections. If there is more than one suitable B-stage matrix, a selection is made. The *path matcher and selector* then sends signals to mark the inlets and outlets associated with the selected links. (In the diagram the extra paths to provide the mark signals are not shown. They are switched by the connector in parallel with the control wires.)

(f) The switches then operate. The control wire must be earthed by either the line unit or the junctors and this earth holds the path. The marking signals can then be removed and the common control is ready to process another connection. The connection can be released by removing the earth on the control wire.

(g) In situations where the *path matcher and selector* cannot find a match, the *control sequencer* can instruct the *junctor selector* to select another junctor (connected to a different C-stage matrix). If no path is found after one or more of these retrials, it is necessary to send a signal to the *line identifier* to ignore that particular calling condition as there is congestion.

It is interesting to note that each of the functional sub-systems in the common control described can be designed as a separate entity. For instance, the *path matcher and selector* has no need to know the number of switch matrices. All it has to do is to make a decision based on the data from the connectors and to transmit a signal to the matrices accessed by the operated connectors. Similarly, the *control sequencer* has no need to know about the detailed operations of the *path matcher and selector* or the *junctor selector*. The only function of the *control sequencer* is to co-ordinate the flow of control signals between the other sub-systems.

Another advantage of the type of organisation described is that it can automatically adjust itself if switches or junctors are taken out of service (or not yet installed). If the control wire code for indicating availability is -50 V, when a matrix or junctor is unequipped the corresponding control wires indicate non-availability, and thus the non-existent (or non-working) units are never selected.

References

1. For instance see Gayford, M. C. (1969) *Modern Relay Techniques*. Newnes-Butterworths.
   and Chapter 2 of Smith, S. F. (1974) *Telephony and Telegraphy A*, Second edn. Oxford University Press.
2. The classic reference is Atkinson, J. (1948) *Automatic Telephony*, Vol. II. Pitman. Chapter 6 discusses the dial pulse reception problem.
3. Keister, W., Ritchie, A. E. and Washburn, S. H. (1951) *The Design of Switching Circuits*. Van Nostrand.
4. Not a spelling error, the translator is named after its inventor T. L. Dimond of Bell Laboratories.
5. For example, Section 3 of Freimanis, L. *et al.* (1966) No. 1 ESS scanner, signal distributor and central pulse distributor, *Bell. Syst. Tech. J.* **63**, pp. 2255–82.
6. See Smith, S. F. op. cit. Chapter 5.
7. Battista *et al.* (1963) Signalling system and receiver for touch-tone calling, *I.E.E.E. Trans. Commun. and Electronics*, **65**, p. 9.
8. The physical and electrical design of the Remreed relay and its uses are covered in a special issue of the *Bell Syst. Tech. J.* (May-June 1976), **55**, No. 5, Remreed switching networks for No. 1 and No. 1A ESS.

10
Practical examples of switching systems

10.1 Introduction

This chapter reviews some practical switching systems. The aim is not to be encyclopedic but to describe examples which are either in common use or illustrative of some particular design philosophy. The chapter deals with electromechanical systems and covers step-by-step, crossbar and reed relay systems. The next chapter covers the area of computer controlled telephone switching systems and Chapter 12 discusses digital switching systems (which are all computer controlled).

10.2 Step-by-step

Strowger. The original automatic switching system was initiated at the end of the last century by A. Strowger and the system still bears his name. In spite of its age, the basic system design remains essentially similar today and Strowger systems are still responsible for switching over half of the world's telephone lines. The reasons for the survival of the design are its flexibility, high system availability, comprehensibility and most important of all, cheapness. Today it is unlikely that many new systems will be installed based on this design, but the long service lives of existing systems will guarantee that Strowger equipment will be needed to extend existing centres well into the 1980s.

The basic principle of the Strowger system is the direct application of functional subdivision with extensive use of third wire control. There is also an element of shared switch network but without any common control. The basic system consists of three sub-systems — pre-selectors, group selectors and final selectors.

A subscriber is connected to a line unit and to a number of final selectors for terminating calls. Within the switching system each speech pair (a- and b-wires) is accompanied by a third, P-wire, for control and signalling purposes.

Some of the essential circuit details at a four digit switching centre are shown in Figure 10.1 [1].

(a) Pre-selector. This is the per-line equipment. It consists of a both-way line unit with its own uniselector which provides the access switch for originating calls. From the idle state the calling condition is detected by the operation of the L relay, which:

— guards the line, that is connects a busy condition to the P-wire of the calling terminal to prevent it being seized by a terminating call,
— maintains the line unit in the busy state until a *clear* signal is detected,
— initiates a search for a trunk to a free group selector.

Contact $L1$ immediately applies a guarding earth to the P-wire to prevent intrusion by any final selector attempting to terminate a call to this particular terminal. The uniselector has access to a number of group selectors whose availability is indicated by the potential on the third wire. A free selector is indicated by -50 V and a busy one by earth potential. Other contacts (not shown) on the L relay start the uniselector stepping around the contacts in order to find the first free selector. Once such a free outlet is found, the K relay operates as it is no longer shunted by the earth via the wiper on the P-bank contact. Contacts of K (not shown) halt the movement of the uniselector. $K1$ and $K2$ disconnect the L relay and connect the terminal directly to the A relay of the first selector. This relay is operated by the subscriber loop and the result is the placing of an earth on the P-wire by the seized group selector. This earth now holds the K relay in the pre-selector. The centre of control is now in the first group selector. The L relay in the line unit is designed to be slow-to-release so that the guard condition is maintained and the K relay is kept operated until the P-wire of the seized first selector becomes earthed. If the user clears before a free first selector is found, relay L releases and it is arranged that this has the effect of returning the uniselector to its home position.

One very important feature of the pre-selector is the provision of two home positions (for example the first and eleventh contact of a 22 position selector). This provides a limited form of random selection since although the search for a free group selector is sequential, the search normally starts from one of two positions. Thus, provided not all the selectors in one half of the group are busy, alternate call attempts pick up different group selectors.

(b) Group selector. The main component of the group selector is a two-motion selector which contains ten levels of ten (or sometimes twenty) sets of contacts. Each group selector has its own set of control relays. It receives a *seize* signal from a pre-selector as a loop condition across its a- and b- wires. This operates the A relay.

This A relay is used for repeating the dialled pulses and hence is balanced to

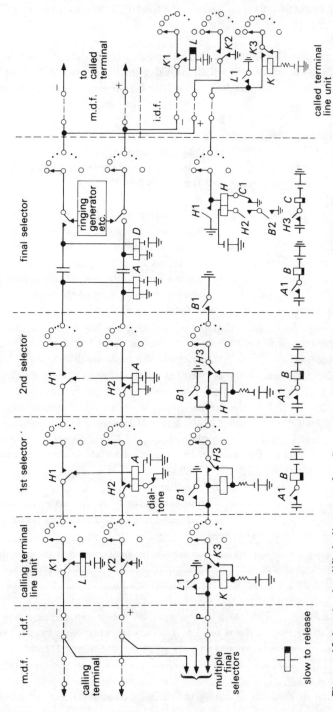

Figure 10.1 A simplified diagram of a Strowger system with a backward holding from the final selector.

the line and designed for high-speed operation. Contact $A1$ operates a second relay, B, which is slugged so that it is slow-to-release. Contact $B1$ applies a guarding earth to the P-wire to maintain the busy condition. This earth provides a hold circuit for the K relay once the $L1$ contact has opened and also maintains the earth on the P-wire of the line circuit via contact $K3$. For satisfactory operation it is necessary to ensure that the release time of the L relay is greater than the sum of the operate times of the A and B relays of the first selector.

The A relay acts as a transmission bridge and is used to return dial-tone to the user. When dialling starts, dial-tone is removed and a contact on the A relay operates the vertical stepping magnet of the two-motion selector to the level corresponding to the first digit. After the inter-train pulse is detected, the horizontal stepping magnet is energised in order to start a search for a second group selector in a manner analogous to the uniselector circuit. When a free outlet is found, the H relay is operated, the user is connected through to the second selector and the A relay in the first selector is disconnected from the line. After a short delay, the B relay of the first selector releases and the line circuit and first selector are held and guarded by the $B1$ contact of the second selector. The centre of control is now in the second selector.

The second selector responds to the second digit and hunts for a free final selector. If the selector is unable to find a free subsequent selector, it is arranged that busy tone is returned.

(c) Final selector. The final selector is stepped into its appropriate position by the third and fourth dialled digits. If the called terminal is free, there is a -50 V potential on the P-wire (via the K relay of the associated line unit). This potential operates the H relay in the final selector via the K relay of the called line unit and this K relay also operates. The effect of the operation of the K relay in the called line unit is to disconnect the L relay and earth from the called line. (This prevents a *seize* signal when the called terminal answers.) In the final selector the H relay is held via contacts $H1$ and $B1$ via a second winding and the K relay in the called line unit is held via the P-wire by contact $H2$. Ringing current is sent to the called terminal from the final selector and ringing tone to the calling terminal. When the called terminal answers by the user going off-hook, this is detected by the ringing circuit which removes the ringing current and connects the two terminals. Feeding current for the called terminal is obtained from a second transmission bridge using relay D.

Charging. It is necessary to transmit the *answer* signal to the calling line unit in order to initiate charging. One way to achieve this is to apply a $+50$ V pulse to the P-wire and thus operate a meter in the calling unit.

Clearing. If the calling terminal goes on-hook, at any stage during the set-up or while the conversation is in progress, the A and B relays of the appropriate selector release and this removes the earths on the P-wires. This removal releases all the selectors and the called terminal. On the other hand, if the called subscriber goes on-hook, the D relay of the final selector releases and produces a line reversal, but in the absence of any other circuitry none of the equipment is released.

This system has *calling party release*; in order to produce *called party release* or *first party release* it would be necessary to have more complicated circuitry, and in most Strowger systems this is not provided for local calls. For long distance working it is often arranged that the trunk is released as soon as either terminal clears. In the example quoted the holding condition is provided from the final selector; the selector train is therefore said to be *backward held* and the called line circuit is said to be *forward held*.

Switch network. The path through a Strowger system is set up on a step-by-step basis. This implies there is unconditional selection of the path, that is a first selector may be chosen which does not then have access to a free second selector. If a different free first selector were chosen by the access switch, access to a free second selector might be possible.

In general the uniselectors connected to a line unit and each level of the two-motion group selectors have access to ten or twenty circuits. The number of sources which have access to these ten or twenty circuits can be decided by a simple application of the Erlang-B equation. For instance twenty circuits can carry 11·1 erlangs at 0·5%, so this will support over 200 sources with 0·05 erlang average originated traffic. When the originating traffic exceeds this level, more servers are required. As the discussion on grading in Chapter 4 shows, greater efficiency of servers is obtained if the traffic is divided into groups and each group has access to a dedicated number of servers, plus an overflow onto a set of servers common to a number of groups.

Register/translators The step-by-step system described above is said to have direct control since the dialled digits directly control the selectors. Figure 10.2 shows how a direct control system can be organised to provide access to other switching systems. In this diagram, each level of selector is represented by a single line. This example also shows how it is possible to construct a switching centre which operates with a mixture of number lengths.

Direct control is economical but inflexible. The discussion in Chapter 5 on telephone network organisation shows that it is desirable in many situations to be able to divorce the dialled number from the actual routing through a network. This can be achieved in a step-by-step system by the addition of sub-systems called *register/translators* (sometimes *directors*). Figure 10.3 shows where they are connected to first selectors.

When a first group selector is seized by a line unit, it in turn seizes a register/translator which returns the dial-tone to the calling terminal. The dialled digits are stored within the register/translator. Once sufficient digits to determine the routing have been received a set of translated digits is generated and

STEP-BY-STEP   237

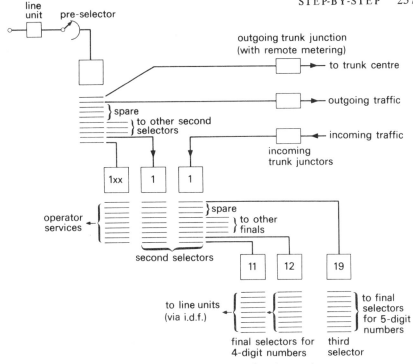

Figure 10.2 Schematic of simple Strowger step-by-step system.

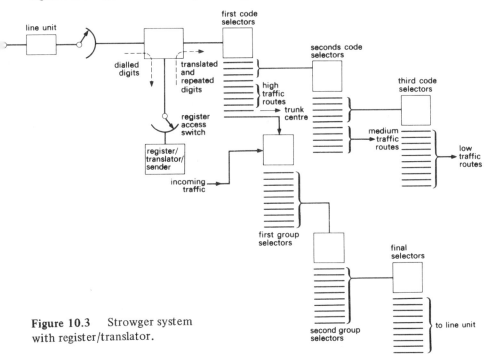

Figure 10.3 Strowger system with register/translator.

238   PRACTICAL EXAMPLES OF SWITCHING SYSTEMS

those digits are fed to the switching train to operate the switching mechanisms. While these translated digits are being transmitted, the terminal normally continues to input further routing digits. The register/translator stores these and retransmits them after the translated digits have been sent.

A register/translator system is said to give indirect control of a step-by-step system. Many technologies have been used for register/translators. The first versions were made with electro-mechanical devices and later versions were in fact the first application of electronics to switching systems [2].

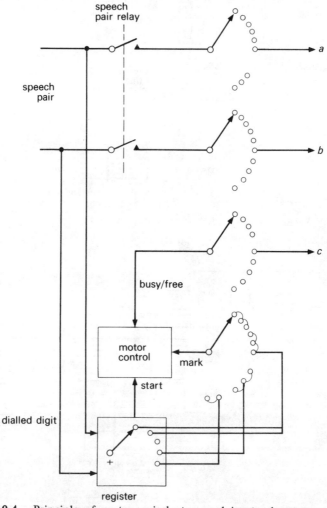

Figure 10.4 Principle of motor uniselector used in step-by-step system. The dialled digit is received by the register which stops a uniselector (or relay equivalent). This uniselector places an earth on a set of contacts corresponding to the possible ongoing circuits for the dialled digit. When the digit is received the motor selector steps over the marked outlets until it finds a control wire which is free. The motor is stopped and the speech path relay operated.

Other forms of step-by-step system. There are many other forms of step-by-step system based on different selector mechanisms. Figure 10.4 shows as an example the German EMD system based on motor driven uniselectors [3]. Each selector contains its own register, which marks the outlets corresponding to the route appropriate to the received digit. This system therefore has a greater flexibility in allocating circuits to codes. The Strowger two-motion group selector can select only 10 routes with up to 10 (or 20) circuits in each route. A register controlled high-speed selector provides up to 500 circuits and these can be allocated to up to 10 routes without restriction.

10.3 Crossbar systems

Ericsson ARF systems. The Ericsson ARF system has a modular design based on a number of sub-systems, each with their individual common-control [4]. A typical system is shown in Figure 10.5. Groups of 1000 subscribers are served by a line switch network controlled by the two markers, for reliability. (Many administrations find that the inherent reliability is high enough for them to omit the second marker.) For originating calls there are two stages of switching and four stages for terminating calls. Each line unit is connected to two identifiers and when a calling condition is detected the line unit seizes (at random) one of the identifiers. The identifier seizes its associated marker once the call has been recognised. Each local junctor provides an availability indicator, showing when it is both free and able to access a free register. The line marker therefore tests these indicators and finds, and operates, a path to connect the calling line unit to a local junctor. The marker is then finished and can be used for a further call.

When a local junctor is seized, it sends a *calling* signal to the register finder control, which proceeds to set up a connection to a register. The register examines the initial digits and, when sufficient have been received to determine the required route, the register calls for a multifrequency sender. The register then instructs the local junctor, to which it is currently connected, to send a *seize* to the trunk switch network.

Each trunk switch network is equipped with its own register and, when a *seize* is received, an inlet identifier (not shown) identifies the inlet and connects the register to the calling inlet. The line switch network register transmits a signal (multifrequency) to the controlling register. On receipt of the *proceed to send* signal, the register instructs its associated sender to transmit the necessary digits needed to route the call out of the trunk switch network. This is normally the first two or three digits of the dialled number.

Next, the trunk switch network register calls the group marker and transfers the received information. The group marker finds an accessible outlet on a suitable free route. If there are no suitable outlets on the first choice route, an alternative route is tried. The trunk switch network register transmits a signal back to the controlling register, to indicate the chosen route (or congestion).

The selected outgoing route transmits a *seize* to the next centre (or to a local

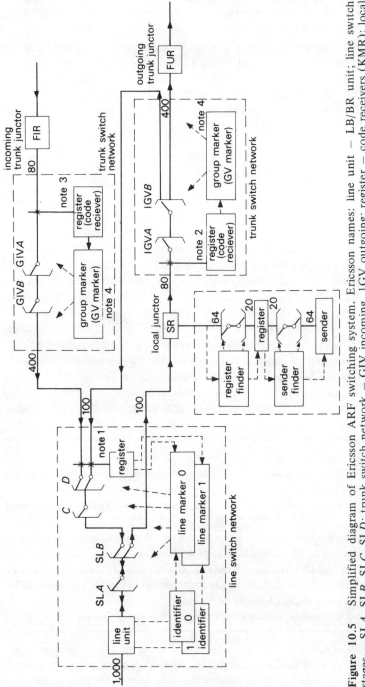

Figure 10.5 Simplified diagram of Ericsson ARF switching system. Ericsson names: line unit – LB/BR unit; line switch stages – SL*A*, SL*B*, SLC, SL*D*; trunk switch network – GIV incoming, 1GV outgoing; register – code receivers (KMR); local junctors – SR; incoming trunk junctor – HFIR; outgoing trunk junctor – FUR. *Note 1*: one register per 100 inlets. *Note 2*: one register per 80 inlets. *Note 3*: normally several registers are required here, so a register finder is needed. *Note 4*: one group marker is shared between two trunk switch networks.

CROSSBAR SYSTEMS 241

line switch network) and subsequently a *proceed to send* signal is returned from that centre to the controlling register. The controlling register can then instruct its sender to transmit the next set of routing digits, and so on.

The controlling register is released once a complete connection has been established. If congestion is detected at any stage, or if an incomplete or invalid number is dialled, the register instructs the local junctor to release the connection. This has the effect of causing busy tone to be sent to the calling terminal, from the line unit. (A typical line unit is shown in Figure 10.6.)

It can be seen that the ARF system maintains a high system availability by virtue of two independent markers in the line switch network and by the

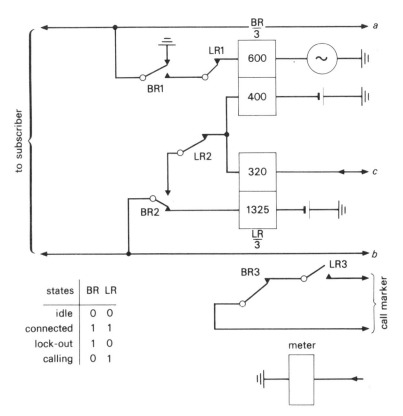

Figure 10.6 Typical line unit for ARF system. In the *idle* state, a loop across the line operates LR and the number is called. When the call is connected to the switch an earth on the C-wire operates BR (and holds LR). The marker call circuit is now released. At the end of a call the earth in the C-wire is removed. This releases LR and if the subscriber has cleared BR is also released. If the subscriber is still off-hook an LR releases and the line unit is in the lock-out state which sends busy tone to the subscriber. To make another call, the subscriber must first clear to restore the circuit to idle. Note that no control equipment is involved in the line lock-out state.

provision of a number of independently controlled sub-systems. Because the controlling register receives backward signals it can usually detect malfunctions in later stages, and make automatic second attempts.

The switch networks are constructed from electrically held crossbar switches with ten verticals and twelve horizontals. With double wiper switching this provides twenty sets of five contacts per vertical. Four of these contacts are used to switch the two speech wires, a control wire, and a billing-meter wire. The fifth contact is associated with a second control wire, used to sense the state of the switches and to operate them.

The capacity of the ARF system can be increased by replacing the register/sender sub-system by electronic systems with micro-processor control. New and modified old installations using this arrangement are designated as ARE systems [5].

A.T. and T. No. 5 crossbar system. The No. 5 crossbar system was developed by the Bell Telephone Laboratories and the first system was brought into service in 1948 [6, 7]. The basic structure is shown in Figure 10.7 with the North American names for the sub-systems used. It consists of two types of switch block called a *line link frame* and a *trunk link frame* (corresponding to line and trunk switch networks). All junctor relay sets and originating registers are on the same side of the trunk link frame. There are a number of markers in a centre, usually between three and twelve, depending upon the traffic. These markers can be connected to a switch network, register, or sender, by means of sets of multi-contact relays called *connectors*. The number of test and operate leads that must be connected to the marker varies between 90 for the line link and 240 for the trunk link frame. There are two types of connector, those which allow an equipment to connect itself to a free marker and those which allow a marker to seize a free equipment. For some equipments both types are required; for instance, the line link frame may call a marker for an originating call and be called for a terminating call.

The sequence of operation is as follows. When a call arises from a terminal connected to a particular line switch network, the control within the network seizes a free marker and passes the identity of the calling terminal to the marker, together with any particular class of service information that may affect the call (for example, coin-box, multifrequency push-button instrument, or long distance-calls-barred). This information is passed by a combination of d.c. conditions on some of the leads and this is decoded and stored on relays in the marker.

At the same time as this is happening, the marker selects an idle trunk switch network which has access to one or more free originating registers. (The trunk switch network is busy if it is being used by one of the other markers for making a connection.) The marker seizes the selected trunk switch network and selects one of the free registers connected to it. The marker then transmits the identity of the calling line to the chosen originating register. Finally, the marker selects and operates a free path between the calling line and the originating register.

CROSSBAR SYSTEMS 243

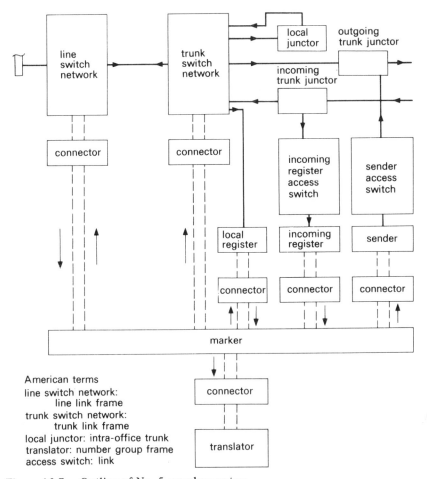

American terms
line switch network:
    line link frame
trunk switch network:
    trunk link frame
local junctor: intra-office trunk
translator: number group frame
access switch: link

Figure 10.7   Outline of No. 5 crossbar system.

Next, the marker releases the line and trunk switch networks and frees itself for another call. Dial-tone is returned from the originating register, which stores the digits as they are received. This complete operation, from calling condition to return of dial-tone takes about half a second (assuming that a free marker is available). At this stage, the centre of control is in the register. This has all the relevant information about the call.

When sufficient dialled digits have been detected by the originating register, the register seizes a marker and passes the dialled information to it. The marker decodes the initial digits to determine whether the call is a local call ('intra-office' in N. American terminology) or an outgoing call. For a local call the marker seizes a trunk switch network which has access to a free local junctor and selects a free junctor. The marker must find which line link frame the called terminal is connected to, and this is achieved by seizing the marker group frame. This is a miniature main distribution frame which is associated with each block of 1000 directory numbers and consists of a relay decoding tree which puts an earth on one of 1000 outlets. These outlets are cross-connected to a corresponding set of inlets which give the physical location of a particular directory number (that is its equipment number) and any relevant class of service information, such as free calls, or incoming calls barred. These terminals are also connected back into the marker via the connector and the application of the earth operates a coded set of relays in the marker. Special arrangements have to be made if the directory number is that of a private branch exchange where there are a number of possible lines corresponding to a single dialled number.

At this stage the marker is able to test the called line via the appropriate line switch network connector and, if the network is free, the marker sets up a path between the calling and called lines via the local junctor. Note that because the register stores the calling line identity it is possible for the marker to find a new path from the calling line to called line. There is no need to connect the call via the path used to the register. The second junctor then controls the call and releases the connection when finished without any necessity to recall the marker.

In all but the smallest centres, there are two types of marker: a dial-tone marker used for the initial connection to a register and a more complex completing marker used for the call set-up.

Trunking arrangement. The trunking of the No. 5 system is based on the use of ten-level crossbar switches with either ten or twenty verticals. These are used

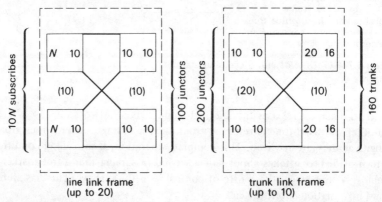

Figure 10.8 Organisation of No. 5 crossbar switch blocks. In line link frame N is typically 29 but can be as high as 59 for low traffic termination. Hence maximum line capacity is around 12 000 lines. Maximum number of trunks is 1000 but some have to be reserved for internal calls. Subscribers are connected to the verticals of the crossbar switches. This saves the necessity of a separate cut-off relay in the line circuit.

as shown in Figure 10.8. The line switch network is a two-stage network with the second stage consisting of ten 10 x 10 switches giving access to 100 junctors. Each first-stage switch connects between 29 and 69 lines on to ten links, which give access to the ten second-stage switches. The number of lines depends upon the traffic; the total capacity of a line switch network is about 33 erlangs (1200 ccs).

The trunk switch network is also a two-stage network, but with less concentration; it connects 200 inlets to 160 outlets. The second stage switches use two-level wiper switching to provide 16 outlets. Since there are 100 outlets from each line switch network and 200 inlets into each trunk switch network, in general the number of trunk switch networks is half the number of line switch networks, with $100/n$ junctors from each switch network (where n is the number of trunk switch networks).

A more recent design of the equipment (called No. 5A crossbar), introduced in 1972, uses a smaller, magnetically latching, crossbar switch, with two-level wiper switching on the line link frames [8]. This increases the capacity and reduces the size of the frames.

Japanese C 400 system. The first commercial version of C 400 was placed into service in Japan in 1965 [9]. It has a similar structure to the No. 5 crossbar system, but uses a different size of switch network (Fig. 10.9). The maximum size is 2800 erlangs and up to 60 000 terminals can be connected (subject to the traffic limitation). Separate markers are used for the dial-tone connection and for call completion.

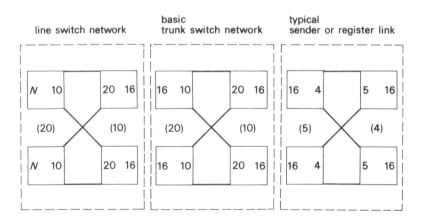

$N = 16 \rightarrow 96$
i.e., $320 \rightarrow 1920$ lines

Figure 10.9 Switch networks used in C 400 system. Basic component is a standard crossbar switch with ten or twenty verticals and ten levels of six-wire contacts. With double wipe switching this provides sixteen outlets.

246 PRACTICAL EXAMPLES OF SWITCHING SYSTEMS

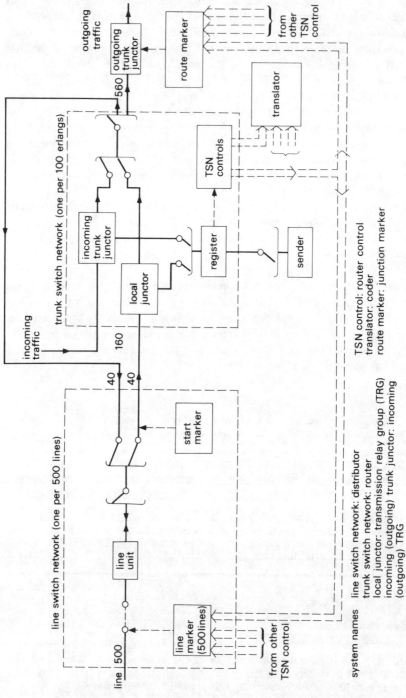

Figure 10.10 Principle of 5005 crossbar system.

A further difference from the No. 5 crossbar system is that C 400 can be adapted to provide register/sender capability. The No. 5 crossbar has to receive a complete dialled code before it can set up an outgoing route and retransmit the code. This is satisfactory for countries which have well determined number lengths, but inconvenient when a network uses a variable length code and when the requirements of international direct dialling are taken into account.

Register/sender working is achieved by the local register calling in a completing marker after the first few digits have been received. If the completing marker decides that the dialled code is for an outgoing route, the marker sets up a connection from the calling line to an appropriate outgoing circuit and connects a register/sender to the outgoing junctor. The digits already dialled are transferred to the register/sender, which then receives the subsequent dialled digits. The register/sender immediately starts transmission of the received digits, so there is a reduction of in post-dialling delay as compared to No. 5 crossbar.

Plessey 5005 crossbar system (British Post Office designation TXK 1 (Local) and TXK 2 international (gateway))

Basic principle. The 5005 system is based on a switch similar to that of the No. 5 system, but the system structure is considerably different [10]. By the use of triple wiper switching it is possible to obtain 28, three-wire outlets from a twelve-level switch. These switches are combined into two-stage line switches and trunk switch networks, as shown in Figure 10.10. The major difference between the 5005 and other systems is that it uses a technique called *self-steering* to set up a path between an inlet and outlet over any number of stages. With this technique, a path is set up by placing an earth potential on a control wire at the point of origin and the point of destination of a connection. Control relays associated with the switches select and connect a suitable free path between the marked inlet and outlet. The use of this technique simplifies the design of the common control elements and reduces the number of wires that have to be brought out of a switch block to a common control.

In a typical local centre the line switches are provided in blocks of 500 lines and these are concentrated onto 40 originating trunks and 40 terminating trunks. These trunks take 20 erlangs each of originating and terminating traffic (an average of 0·04 erlang originating traffic per terminal). Higher average traffic levels can be accommodated by adding up to eight extra second-rank switches in parallel with the eight basic ones.

The originating traffic from the line switch, together with incoming traffic from other centres, is connected via junctors to the trunk switch network, which provides a total of 560 outlets. Each trunk switch network can cater for about 100 erlangs. The trunk switch network contains the overall control of a call and either routes it back to the correct 500-line switch network for a local call or to a suitable outgoing trunk circuit.

Local call operation. The sequence of operation for a local call is that the calling condition is detected by a line circuit and this circuit applies a marking potential to its associated inlet. A relay contact chain is provided to ensure that only one call applies a mark if two or more calls arise simultaneously on the same 500-line switch network. All free junctors from the trunk switch network which have access to a free register are also marked. Each trunk switch network has a start marker which selects and marks one of the free junctors. This operates the path between the line unit of the calling line and a local junctor in the trunk switch network. A cut-off relay in the line unit operates, and this removes the marking condition and connects the calling line to the local junctor. The start marker is arranged to step around the free trunks after each call in such a way that the traffic is distributed evenly over all the trunk switch networks. During the path completion operation, the local junctor seizes a register and, once the line is connected, the register returns dial-tone; the user can then enter the number required.

If the register recognises a local call, it stores the required number of digits and then seizes the trunk switch network control to complete the connection. This control seizes the appropriate 500-line switch network marker, which consists of a relay tree with five 'hundred' relays and 50 'ten' relays. Operation of the appropriate relays connects the control wire of the called line unit to the trunk switch network control. The condition on this wire indicates the state of the called line.

When the line is busy the associated local junctor it is switched to sends back 'busy'. If the line is found to be free, it is marked by placing an earth on its associated inlet and at the same time the control places an earth, via the register, onto the outgoing side of the local junctor; the self-steering circuit then connects the path between the local junctor and the called line. The register is disconnected and the call is supervised by the local junctor, which releases the connection at the end of the conversation without any further need to use common equipment.

In the 500-line units there is an individual intermediate distribution frame which allows any of the 500-line units to be connected to any of the 500 marking leads. So, any line circuit may have any directory number within the block of 500.

Outgoing call. For an outgoing call the required route is decided by the first few dialled digits. The register seizes a translator which provides the route number and other information, such as the number of further digits expected and the change rate. The control seizes an outgoing route marker which marks all available circuits to the selected route. The path between the local junctor and the outgoing trunk junctor is then completed by the self-steering mechanism. When a path to the next centre is established, a sender is connected to the set-up to transmit the remainder of the dialled numbers.

CROSSBAR SYSTEMS

System security philosophy. The essential common control in the line switch network is effectively duplicated by the use of double wound relays and two sets of contacts. The trunk switch networks contain in fact two controls and the switches are divided into two physically separate blocks, each block associated with one of the controls and each block having access to separate sets of outgoing routes. The first attempt at setting up a call is made with either router control, but if a second attempt is necessary it will always use the alternative router control. The second attempt may be necessary because no free path is found or because the path fails a continuity check once it has been set up.

Pentaconta system (British Post Office designation: TXK 3 (local) and TXK 4 (trunk).) An example of a crossbar system designed around a different size switch is that of the Pentaconta system which is based on a switch having 22 verticals and 14 select bars [11]. This provides 28 output levels or 52 with double-wiper switching. With triple-wiper switching the number of levels may be increased to a total of 74.

A typical arrangement of a line switch network is shown in Figure 10.11. This is basically a two-stage network with three links between each *A* and *B* switch. These links are arranged so that there is one individual link between two particular *A* and *B* switches and two links from one *A* switch, each of which is common to two *B* switches. The traffic capacity is increased by what is called an *interaid* network (or *entraide* in French) which provides a third stage of switching if the two-stage network is blocked. If an incoming call cannot reach the required line via one of the three available links, it is switched on to one of the interaid links, which can gain access to a *B* switch that does have access to a suitable free link. If this fails then the preceding group selector is instructed to make a second attempt on a different primary section. In the configuration

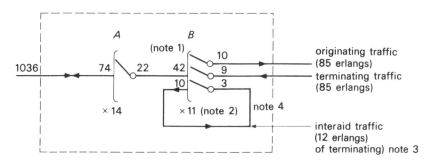

Figure 10.11 Line switch network using interaid.

Notes: 1. The links between *A* and *B* stages are arranged such that there is one link between each *A* and *B* plus one link from each *A* to two *B* stages. Some inlets on *B* stages therefore receive links from two *A* stages. 2. For lower traffic requirements, fewer *B* switches are provided. 3. A small amount of originating traffic (~0.15 erlangs) also uses interaid network. 4. Interaid links typically arranged so that each link has parallel access to the three to four other *B*-stage switches.

shown in Figure 10.11 about 14% of the incoming traffic is taken on the interaid links.

One advantage of the interaid system is that it is not necessary to apply conditional selection over the complete switching centre, since there is a low probability that an incoming call on a particular B switch inlet will be blocked. This in turn means that the system can be designed in self-contained units, which makes the system applicable to a wide range of centre sizes.

10.4 Reed relay systems

TXE 2. Reed relay systems have many similarities to crossbar systems in their layout and control. The main difference is that they have more stages of smaller size.

Thus a reed relay switch network is near-optimal in crosspoint efficiency. Crossbar systems, on the other hand, minimise cost by cutting the *actuator* cost per erlang, rather than crosspoint cost per erlang. For instance, the TEX 2 small exchange system (200 to 4000 lines) in common use in the United Kingdom uses matrices of 5 × 5 and 5 × 4 crosspoints. [12] Each crosspoint consists of a 4-reed insert in a single unit. An outline of the complete system is shown in Figure 10.12. It consists of a three-stage line switch network, together with a single-stage trunk switch network. Each A switch consists of twenty-five 5 × 5 matrices which between them connect 125 lines on to 25 links between the A and B stages. The 25 links are arranged as shown in Figure 10.13. No two 5 × 5 matrices have more than one link in common. This arrangement results in good traffic handling capacity and permits a random mixture of low and high calling terminals on each 125-line block. Traffic simulation studies show that the traffic carrying capacity on the 25 $A-B$ links is 7·5 erlangs for a grade of service of 1%.

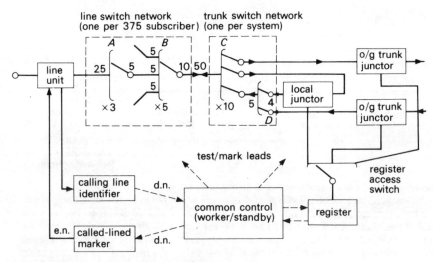

Figure 10.12   TEX 2 reed relay system.

REED RELAY SYSTEMS 251

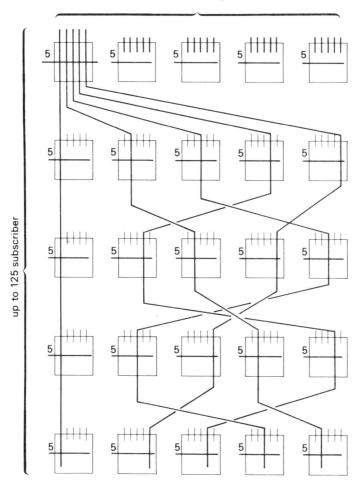

Figure 10.13 Internal arrangement of *A* switch for TXE 2.

There are two common-control equipments, one of which is operational at any one time. Control is switched between the two at regular intervals or whenever a fault is detected in the active control. When a user initiates a call, a contact of one of the line unit relays sends a pulse of current into the calling line identifier. This is a Dimond ring store, consisting of four rows of ten large pulse transformers, each row representing the thousands, hundreds, tens and units part of the user's directory number. A wire from each line unit is threaded through the appropriate transformer in each row and there are secondary windings on each transformer connected to sense circuits. So the pulse of current produces the directory number of a calling terminal. This number is stored in a free register which then seizes the common control and causes it to set up a path between the calling line unit and a free outgoing trunk, and between this junctor and itself. The register then returns dial-tone. This complete operation takes only 60 ms. The user can then dial his number.

If the register detects that the call is a local call it seizes the common control which changes the path over to a local junctor. Because of the high speed of the

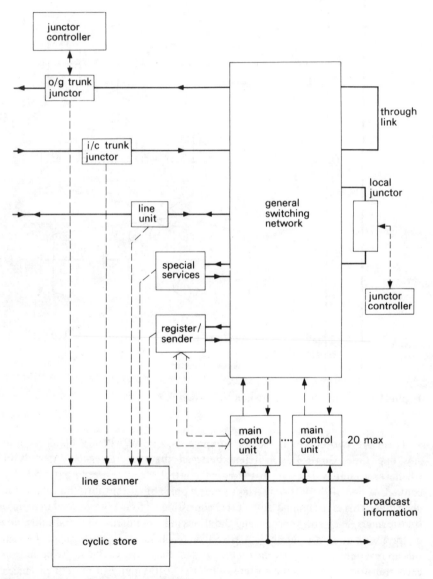

Figure 10.14 General view of TXE 4.

REED RELAY SYSTEMS   253

control and the path switching this can be achieved during an inter-digit pause. The system may therefore be used without any senders. This is exploited in the U.K. to minimise costs, but it is found to be too inflexible for the introduction of linked numbering within numbering plan areas.

TXE 4. TXE 4 is a reed relay system for large centres (2000 to 40 000 connections or 5000 erlangs) [13]. The first system was brought into service in 1976. Figure 10.14 gives a general view of the TXE 4. The main point of note is that it uses a single general purpose switching network. On one side of the network are connected all local, lines, incoming and outgoing trunks, junctors, registers, senders, and any special equipment (such as the unit for controlling a coin box). The other side of the network has only internal junctors. These are of two types:

— through links which provide a metallic path,
— local junctors which provide a transmission bridge.

Both types can break the speech path under instruction from the control.

The use of a single network permits the interconnection of any devices on the traffic side of the switch and provides considerable simplification in control of the system and in the provision of extra services. This technique is called *serial trunking*. Each through link or local junctor can be connected to any terminations on the traffic side by three stages of switches on one port, and four stages on the other port. So, a simple connection from one termination to another uses seven stages of switching plus link.

The use of serial trunking is shown in Figure 10.15. A new call is first connected to a register which receives the digits. For a local call a parallel path is established from the calling line to the called line via a local junctor. The local junctor provides a ringing current control, clear supervision and path release.

For an outgoing call, each register has an associated sender which is connected to the traffic side as well. Once the outgoing route has been determined, the sender part of the register sender is connected to an outgoing trunk junctor (Figure 10.15b) and sending commences. At the same time a speech path (via a through link) is established directly between the calling line and the selected outgoing trunk junctor. In this case the line split relay of the through link is operated. Once the register/sender checks that the connection is established, the paths connecting the register/sender are released and the through link line split is cancelled. The connection is then under the control of the outgoing trunk junctor.

Many more complex sets of paths may be established as required.

Switch sub-system. One overriding design consideration of TXE 4 was to achieve high levels of system availability. One major technique by which this is achieved is the use of separate 'planes' of switches as shown in Figure 10.16.

254 PRACTICAL EXAMPLES OF SWITCHING SYSTEMS

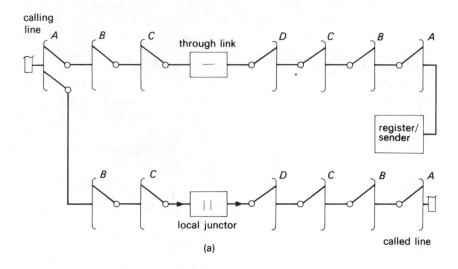

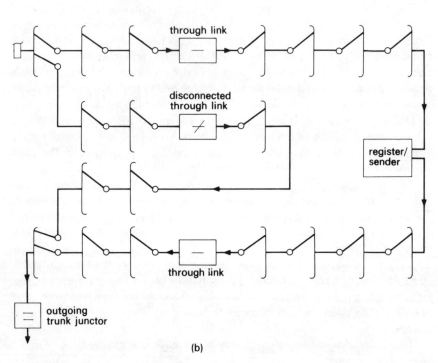

Figure 10.15 Connections in TXE 4 showing use of serial trunking. (a) Local call. (b) Outgoing call.

REED RELAY SYSTEMS 255

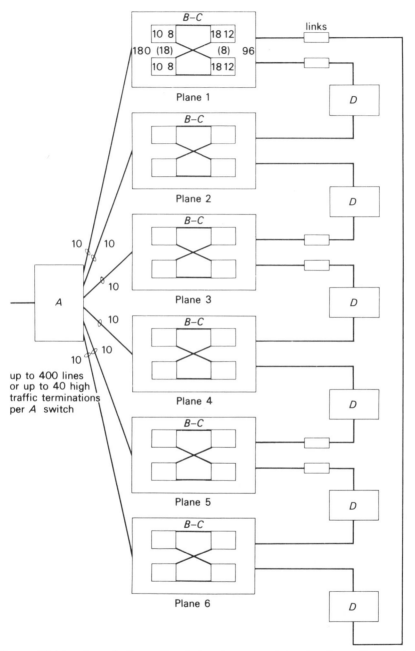

Figure 10.16 Organisation of switch subsystems into six planes (larger size centres use eight planes).

There are six (or in larger systems eight) separate switch blocks containing the B and C stage switches. Groups of up to 400 lines or up to 40 high traffic terminations are connected to an A switch, which gives each termination access to each of the six (or eight) planes. So, any fault on one of the planes will affect only the traffic capacity of the system.

Each plane consists of a number of two-stage switch matrices connecting 180 inlets to 96 outlets. The number of such sub-units depends upon the size of the system and can be up to 48. A sub-unit of one plane is connected to all the sub-units of adjacent planes (via a link and a D switch). The D switch is a square matrix whose size is given by the number of sub-units in a plane.

Each inlet on an A switch has access to the six outlets connecting it to the six planes (three to odd-numbered planes and three to even-numbered planes). It is arranged that no two groups of four inlets have access to the same set of B stage inlets. The number of rows of 4 x 6 matrices is adjusted according to the traffic. For an A switch used for high traffic items, only the first row is provided. This assembly allows up to 400 terminations access to 60 outlets. Details of the path search and operations can be found in Reference [13].

Control. Control of TXE 4 is vested in a number of independent main control units (up to a maximum of 20). Each of these units has a number of register/senders (typically 36 each). One of the problems of a system with a number of independent controls is co-ordinating the action of the individual controls in such a way that a fault or malfunction in one does not affect the whole system. This problem is overcome in an ingenious way on TXE 4. The system is equipped with a number of (duplicated) cyclic scanners which periodically broadcast all the fixed information relating to the system. This information includes the association between directory and equipment numbers, between access codes and specific outgoing trunk junctors, and so on. At the same time a line scanner (again duplicated) broadcasts the state of line of the various high traffic items.

The rate of broadcast of individual items depends upon their function. The slowest scan rate is at periods of 156 ms for local lines. The states of most other equipment are broadcast every 36 ms, with the exception of incoming trunk junctors from step-by-step systems which are scanned every 12 ms.

The sequence of operation is as follows. A user initiates a call and within 156 ms the state of the associated line unit is broadcast, together with the information from the cyclic store giving e.n., d.n. and any relevant class of service details (such as multifrequency keyphone terminal). At any one time one of the free main control units is designated as looking for the next call. When this designated main control unit detects a calling condition, it instructs the switch network to connect the calling line to one of the registers controlled by the main control unit.

The register detects when a significant number of digits has been dialled and itself calls in the main control unit by indicating a change of state to the scanner.

For an outgoing call the main control unit then compares the received code with the broadcast information and looks for the concurrence of dialled code and free outgoing trunk junctor. The main control unit selects that junctor and instructs the switch network to set up the required paths.

Once a connection has been established, and the register/sender has been released, the connection is under the control of the local (or outgoing) junctor. The logic circuits in these junctors are minimal and all the logic is provided by a special purpose time-shared processor. One such processor can control up to 3000 junctors, and is triplicated for reliability.

The first generation of TEX 4 uses a Dimond ring arrangement for the cyclic store and for the program store in the main control unit. In later versions (called TEX 4A) these are replaced by large scale integrated circuit devices.

References

1. Circuit details may be found in Atkinson J. (1974) *Telephony*, Vol. II. Pitman, or Smith, S. F. (1974) *Telephony and Telegraphy A*. Oxford University Press.
2. Heron, K. M et al. (1951) An experimental electronic director, *Post Office Elect. Eng. J.*, 44, pp. 96.
3. Trautman, K. (1967) *Design of Automatic Telephone Exchanges*, Vol. I: *Circuit Units and Functions*, Siemens Aktiengesellschaft.
4. A good review of circuit principles involved in the ARF system may be found in Freeman, A. H. (1973) *Automatic Telephony in the Australian Post Office*, Australian Telecommunication Monograph No. 4 Telecomm. Soc. Aust.
5. Meurling, J. and Eriksson, R. (1974) ARE – a dual purpose switching system for new installations and for updating exchanges in service, *International Switching Symposium*, paper 445.
6. Korn, F. A. and Ferguson, J. G. (1950) The No. 5 crossbar dial telephone switching system, *Trans. American Institute of Elec. Eng.*, 69, pp. 244–54.
7. Graupner, W. B. (1949) Trunking plan for No. 5 crossbar, *Bell Lab. Record*, 27, pp. 360–5.
8. Holtfreter, R. P. (1970) A switch to smaller switches, *Bell Lab. Record*, 48, 2, pp. 46–50.
   Smith, F. H. and Catterall, J. M. (1971) Uses for the new small crossbar switch, *Bell Lab. Record*, 49, 7, pp. 216–21.
9. Tashiro, J. (1966) C 400 type crossbar switching system, *Japan Telecomms. Rev.*, 8, 4. Nagata, T. Miyazu, J. (1967) C 400 crossbar automatic exchange, *Japan Telecomms. Rev.*, 9, pp. 173–81.
10. Corner, A. C. (1966) The 5005 crossbar telephone exchange switching system Part 1 – trunking and facilities, *Post Office Elect. Eng. J.*, 59, pp. 170–7.
11. Basset, J. P. and Camus, P. (1963), 1000 B Pentaconta crossbar switching system, *Elect. Comm.*, 38m pp. 196–212.
    Bernard, R. et al. (1968) Subscribers stages in Pentaconta PC 100 C exchanges, *Elect. Comm.*, 43, pp. 346–54.
12. Long, R. C. and Gorringe, G. E. (1969) Electronic telephone exchanges TXE 2 – a small electronic exchange system, *Post Office Elect. Eng. J.*, 62, pp. 12–20.
13. TXE 4 Electronic exchange systems
    Part 1 – Overall description and general operation, Goodman, J. V. and Phillips, J. C., *Post Office Elect. Eng. J.* 68, pp. 196–203 (Jan. 1976).
    Part 2 – Design of switching and control equipment, Phillips, J. C. and Rowe, M. T., *Post Office Elect. Eng. J.* 69, pp. 68–78 (July 1976).
    Part 3 – System security and maintenance features, Huggins, G. et al., *Post Office Elect. Eng.* 70, pp. 12–20 (Apr. 77).

11
Computer controlled switching systems

11.1 Software organisation of computer controlled centres

The general principle of time-divided control is described in Chapter 3. When a computer is used to control a number of sub-systems (such as registers or switch networks), the computer is effectively being multiplexed over a number of control units. There are three classes of technique whereby this multiplexing can be achieved:

> time division,
> function division,
> call division.

This section is written on the assumption that the reader is familiar with computer organisation and general programming techniques. The discussion in this section is based on the simple system shown in Figure 11.1a with the line unit and junctor shown in Figure 11.1b and 11.1c. More detailed discussions are found in the references [1] and the second half of the chapter discusses some aspects of practical systems.

Time division. The time-division technique is that outlined in Section 3.6. With this technique a processor offers service to a number of control units in turn. This is called *scanning*. The processor performs any necessary actions as it detects changes in the input conditions presented via the input interface. The processor moves on to the next unit after it has performed the required actions for a particular unit (including updating the state memory of the control unit). After dealing with the last control unit, the processor waits for a clock interrupt to restart a scan of the control units. The use of a clock interrupt ensures that the average rate of scanning a particular control unit is fixed and therefore timing operations can be done by counting. For this technique to be successful it is necessary to ensure that

SOFTWARE ORGANISATION 259

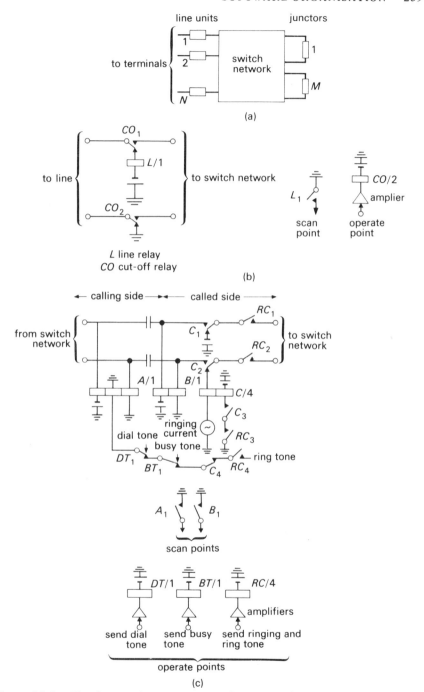

Figure 11.1 Simple stored program control system. (a) General view; (b) line unit; (c) local junctor.

(a) there is a sufficiently low probability of the scanning overrunning a scan period. An overrun could be caused by an abnormal number of calling conditions occurring in one scan period.
(b) if an overrun does occur, its effect is limited, and in any case, will not then cause the program to malfunction permanently. The effect of an overrun might be to miss some event. This may cause a call to be misrouted or, if the overrun persists, control units situated late in the scan chain to be denied service. In either case, the program continues to function normally once the overload disappears.

The simplest time-division technique is to scan all the line units and then scan all the local junctors. When the processor detects a calling condition on one of the line units it has to select a junctor and select and operate a path through a switch network.

Selection of junctor. There are two techniques for selecting a free junctor. In the first, one bit per junctor is allocated in a data area in the computer memory and this bit is set to 'one' if the junctor is free. The set-up program looks at these bits, selects a free junctor and immediately sets the associated bit to zero. This is known as the '*map-in-memory*' technique.

Alternatively, the '*map-in-network*' technique can be used. In this technique the junctor itself contains a one bit memory element, which is read by the path set-up program to check whether it is free. This technique takes more time and increases the cost of the junctor, but it is advantageous if there are several processors controlling the system, because it simplifies the concurrency problem (several processors simultaneously trying to seize the same junctor). A second practical advantage is that, if a junctor is taken out of service (or has never been provided), the scanning process can automatically return a busy indication. In this case the program may be written as though all junctors are working; it will seize only those which are available (that is the same approach as is used for electro-mechanical systems).

Selection of path Once the junctor has been selected, a path must be found between the calling terminal and selected junctor. As with the selection of a junctor, a 'map-in-memory' or 'map-in-network' technique may be used. However, a network is so large that providing one bit for every possible interconnection would be uneconomical. Instead, only the busy/free state of each link between the stages is stored (in the network or processor memory). To see how this works, consider the three-stage network of Figure 11.2, where the internal structure of each switch is shown. Assume that there is a new calling line on switch 1 of the A stage and it is to be connected to the calling side of a selected junctor on switch 2 of the C stage. Typically, all the busy/free bits for the links from a particular A stage switch would be stored in one word in the control memory and all the busy/free bits for the links to a particular C stage

SOFTWARE ORGANISATION

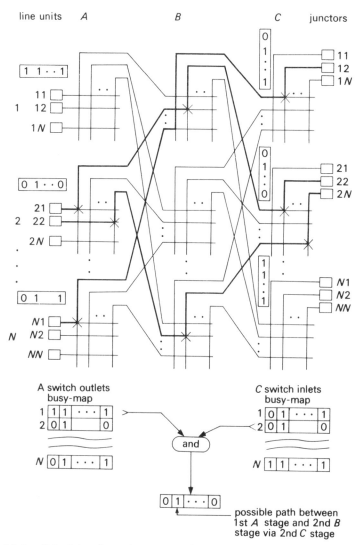

Figure 11.2 Principle of map-in-memory for speech path control.

switch would be stored in another word. The possible paths between the line unit and junctor may be found by performing a logical AND operation between the word corresponding to the map of the links from switch 1 in the A stage, and the word corresponding to the map of the links to switch 2 in the C stage. The bits that remain at 'one' after this operation indicate those B stage switches through which paths are possible. (For the example shown in Figure 11.2, the only free path shown corresponds to that using the second B-stage switch.) The path-search program can then select one particular B-stage switch, set the appropriate bits to busy, and issue commands to operate the switches to set up the selected path. If there is no free path available, another junctor may be chosen and a second attempt made. In a multistage network the algorithm becomes more complex [2].

Storing only the busy/free states of the link produces a significant reduction in the memory requirements. However, it does introduce ambiguity when the path is cleared. This ambiguity occurs when there is more than one connection through a particular switch. All that is known is which inlets and which outlets are in use. As may be seen from Figure 11.2, the state of the links does not indicate which path to clear if the connection between line unit 21 and junctor 22 is to be cleared. For this reason, a map-in-memory system needs to store the actual path for each connection. (In this simple case, the path is completely specified by the second-stage switch used.)

When the path through a switch network is held by a potential on a control wire the path can be cleared by changing this potential. In this case it is always necessary to use the map-in-network technique to find the path for a new connection.

Path operation. Only one path can be set up at a time. If electro-mechanical elements are used, the operating time is from 1 to 100 ms. Since it is possible for several line units to call simultaneously, the set-up program must first check that a path operation is not already in progress. If one is in progress, the set-up program must wait until it is finished. This waiting is easy to implement in our simple time-division system, because all that is required is a single bit indicating whether a path operation is in fact in progress. As soon as the set-up program requires a path, it checks this bit and sets it to busy if it is free. If the bit already indicates busy, the set-up program does nothing but checks again on the next scan. If two or more line units are waiting to set up a call this implementation implies a priority system based on the scanning order.

The path-operation program may also run on a regular basis. This means it can issue an instruction to the distributor to operate the appropriate relays, and then wait a predetermined number of scans to allow them to operate. Alternatively (and preferably), once the relay has operated it can give some indication, that can be checked each scan, of whether the operation was successful. If no success is noted after a certain number of scans, an error is assumed and a second attempt is made using a different path.

Once a path has been successfuly set up between the line unit and the calling half of the junctor, the main program issues an instruction to operate the cut-off relay of the calling line unit and the dial-tone relay in the selected junctor. After a short period, the calling-line relay in the junctor is operated by the loop from the terminal being switched through. The next task is to detect the dialling. On each scan of a line unit, the program looks at the A relay contact of the junctor to which it is now connected. The first change expected is from make to break.

When this occurs it indicates either the start of dialling or else a premature release. The dial-tone relay is instructed to release, and the time of the break period is measured to distinguish between a dialling break (of the order of 66 ms) and a *clear* signal (greater than 200 ms). With a 10 ms scan period this distinction is possible by counting the number of scans for which no change occurs. If this number exceeds 20, the break condition can be assumed to be the *clear* signal. The duration of a make must also be measured to distinguish between the 33 ms makes within a dialled digit or the much longer (minimum 220 ms) inter-digit pause. This too can be done by counting.

When sufficient digits have been received to determine the required terminal, the program tests the busy/free state of the called terminal. If it is busy, the *BT* relay is operated. On the other hand, if it is free, it is busied and a path is found and connected as in the initial path set-up. The cut-off relay of the called line unit is operated and *RC* in the junctor. In the connected state, the cut-off relays of the line unit are operated, so the loop condition of the two terminals involved is only detectable at the junctor. The program checks the line condition of the associated junctor on each scan to detect either an *answer* condition, or a *clear*.

Function division. The simple time-division scheme scans all units in the same period. This has the disadvantage that the scanning must be performed at the fastest rate. In our example the maximum period between scans must be 10 ms in order to detect the dial pulse. However, the line units need only be scanned every 100–200 ms and during the supervisory phase the junctors need only be scanned about every 100 ms. Thus, for most of the time, the scanning has little to do but occupies the processor which might otherwise be performing other work.

The real-time occupancy of the processor can be improved by functional division of the software. The program is split into several modules, each of which is activated at a rate suitable for its operation. These modules are called *processes*.

For instance, in the above example, the program could be split into:

> line scanning (100 ms rate);
> call set-up (whenever a new call is detected);
> initial path set-up (say 20 ms rate);
> dialling (10 ms rate);
> final path set-up (20 ms);
> supervision (say 100 ms);
> path clear down (20 ms).

Now, instead of each control unit being scanned at 10 ms intervals to see whether it needs service, each function is activated at its appropriate rate and is supplied with a list of addresses of the control units needing its particular service. The situation therefore resembles a factory production line where a call is passed from function to function as needed.

There are many ways in which the list of line units needing service can be stored. While a terminal is idle, the only storage needed is 1 or 2 bits to indicate its state. These are called *line records*. A large number of words is required while a call is active (typically 4–16 16-bit words). This block of storage area is commonly called a *call record*. The number of call records that need to be provided is only equal to the maximum number of simultaneous calls that are expected.

Each function needs a work list indicating which call records require its particular service. Techniques for this are discussed in Reference [1]. When a process determines that one of its call records no longer needs its services, the process must arrange to transfer that call record to the work list of the next process. This is shown diagramatically in Figure 11.3. The work list of the line-scan process consists of all the idle call records.

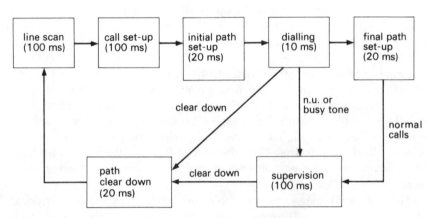

Figure 11.3 Interrelationship of functionally divided processes.

In a functionally divided system, it is necessary for the processor's time to be allocated to the various functions. This time allocation is performed by a program given many different names in different systems, such as 'operating system', 'executive', or 'scheduler'. Here we call it the executive. The executive uses a sub-program, which will be called the scheduler, to decide on priorities. The executive must arrange to activate the various processes at their appropriate rate. Since the fastest rate corresponds to a scanning period of 10 ms, a clock interrupt is needed to initiate the process. At each interrupt the 10 ms processes must be activated; at every other interrupt the 20 ms processes must be activated, and so on. See References [1] and [3] for details.

In the approach described above the line-scan process scans each line unit separately and uses 1 or 2 bits in memory to store the state of the line. This is inefficient, both in program running time and computer memory, as the probability of a line calling in one scan period is very low. For instance, a typical

SOFTWARE ORGANISATION

high-traffic terminal might originate an average of only one to two calls in the busiest hour. Hence the probability of a new call arising in a 100 ms interval is about 1 in 20 000. The amount of time spent scanning can be reduced considerably by scanning lines in parallel. For instance, if the control processor has a 16 bit word size, it is convenient to scan 16 lines with one instruction yielding a 16 bit word A (Fig. 11.4). Within the processor there is a 16 bit word for each of the groups of 16 lines and this word represents the line unit condition read at the last scan. The scanning process need only perform an exclusive OR on these two words. If the result is zero, no change has occurred, nothing needs to be done, and the next group of 16 lines can be scanned.

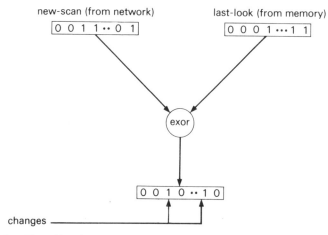

Figure 11.4 Use of last-look with parallel scanning.

However, if the result is not zero, the position of any non-zero bits indicates which of the lines within the group of 16 has changed. The scan process places the identity of the calling line unit plus the direction change in a queue to be dealt with by the call set-up process. The scan program also updates the last look word in the memory.

Similar techniques may be used for detecting other events such as counting dial pulses. Sixteen junctors may be scanned in parallel. Only those for which a change has occurred are processed.

Call division. Function-division systems work well, but require great care in their implementation because:

(a) Each process is responsible for deciding when to transfer a call record to another process, and this often implies that the decision-making points of the call-processing program are distributed over many different parts of the program. This makes it more difficult to debug and to modify the system.

(b) There are many work lists with processes at different rates, so it is difficult to detect the onset of processor overload and thus initiate some load-shedding action.

A software structure which overcomes these difficulties is that of call division. In this technique there is effectively only one process for processing telephone calls and, whenever an event occurs, this event, plus the current state of the call, are presented to the call processor for action.

The basic principle of call division is shown in Figure 11.5. This resembles the basic control system organisation of Figure 3.13. In this structure the event detector processes are scanning the speech path sub-system. In general, these programs are functionally divided, as described above. When an event is detected, the event detector processes place a two-word entry in the input queue. One word gives the identity of the item of equipment on which the event was detected; the other word is a code for the particular event found. The event detectors are triggered by the scheduler, and their only action is to place the items in the queue.

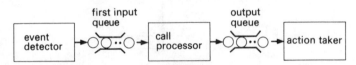

Figure 11.5 Organisation of a call-division system.

The call processor takes the first item from the input queue and, by examination of the equipment identity, finds the call record which is dealing with the particular call. If the equipment identity is that of a line unit, the call processor obtains a new call record and initialises it. The call record contains the current state of the call and this, plus the new event, determines the action to be performed and the new state to be stored. It should be noted that this information can be deduced directly from the state transition diagram.

There are three ways to implement the actual decision making:

(a) A routine may be selected on the basis of the current state and then called. The routine examines the new event plus any other relevant information, performs any necessary input/output operations and loads the new state into the call record.

(b) A routine may be selected on the basis of the event and then called. The routine examines the current state to determine its precise course of action.

(c) The current state and the new event may be used as parameters to address a look-up table. This gives the new state plus the identity of a routine to move the call to this new state.

All three techniques are used in practice. The first is preferable when there are few states and many possible events, and the second when there are few events and many states. The third technique is advantageous when there are many states and many possible events, or when it is required to centralise the actual decision-making logic.

Look-up tables. The look-up table to implement the third method could be a simple two-dimensional array but

(a) in general, not all events can occur for all states and therefore the array is likely to contain many gaps and be inefficient in the use of storage space;
(b) the next state is not always uniquely determined by the new event and the current state (as stored in the call record).

Examples of the latter occur when there is some form of blocking, for instance, if no path or junctor is available to a calling line. In this case there could be several possible next states. The actual next state depends on circumstances external to the call record. What this means is that the state as stored in call record is not complete, but should take into account the states of everything else in the system. This leads to an astronomical number of states and therefore the state of the rest of the system is only investigated when necessary.

The effect of the rest of the system may be taken into account by calling a sub-routine to check the system; this sub-routine returns with a value which can be interpreted as a pseudo event.

11.2 No. 1 ESS

The first production model of a computer controlled switching centre was developed by Bell Laboratories and went into public service in 1965 [4]. The original design had as an objective a maximum of 65 000 lines and a maximum calling rate of 25 000 calls in the busy hour (about 1500 erlangs). Improvements in design and technology have over the years increased the capacity of the system. Systems going into service in the late 1970s had a capacity of up to 130 000 lines and 240 000 calls in the busy hour (equivalent to about 10 000 erlangs).

No. 1 ESS consists of two sub-systems (Fig. 11.6):

— peripheral sub-systems containing the switch networks, junctors, senders and receivers;
— processor sub-systems containing duplicated processors with call and program memories.

268 COMPUTER CONTROLLED SWITCHING SYSTEMS

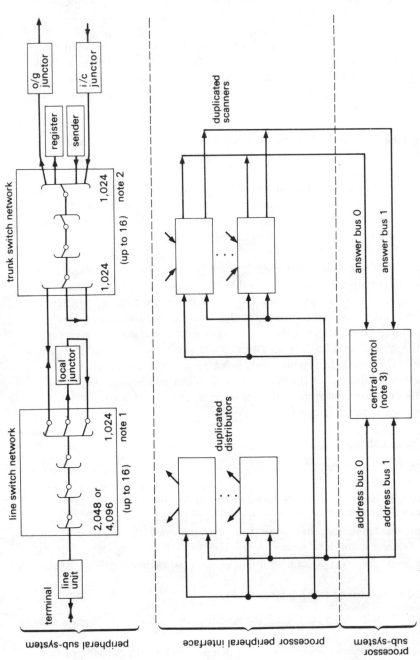

Figure 11.6 General view of Bell No. 1 ESS. Bell names: line (trunk) switch network – line (trunk) link frame; local junctor – junctor circuit; i/c (o/g) trunk junctor – i/c (o/g) trunk circuit. *Note 1*: in No. 1A ESS a maximum of 28 l.s.n.'s are possible. *Note 2*: in No. 1A ESS the t.s.n.'s are 2048 by 2048. *Note 3*: in No. 1A ESS the central processor is based on 1A processor.

Peripheral sub-system. The switch networks in the original version are built from the magnetically latching Ferreed (see Ch. 2). A line switch network consists of four stages, concentrating 4096 lines (or only 2048 in high traffic centres) on to 1024 internal junctors [2]. Trunk switch networks are also four stage, connecting 1024 inlets to 1024 outlets. In No. 1 ESS there can be up to 16 line switch networks and up to 16 trunk switch networks. The actual maximum is usually limited by processing capacity rather than by network

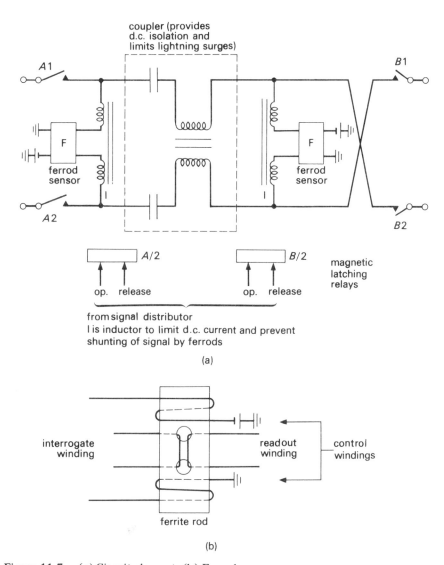

Figure 11.7 (a) Circuit element; (b) Ferrod sensor.

design. In 1976 a new design of switch network was introduced, based on the smaller magnetically latching Remreed [5]. These new networks are plug compatible with the old networks so that they can be intermixed with the old design.

The junctor circuits are designed to contain the minimum of hardware required to provide the necessary power and tone feeding conditions, and the necessary signalling conditions to the lines and trunks.

All logical operations are performed by the central control. This includes determining the number dialled; the control scans the dial-pulse registers at 10 ms intervals. This represents a significant load for the processor and in 1968 an additional (duplicated) processor was added to perform the routine input/output logic operations in the larger switching centres.

As an example Figure 11.7a shows the elements of the local junctor. This contains on each side a feeding bridge and sensing elements called *ferrods* (Fig. 11.7b). There are two magnetically latching relays which connect each side of the junctor to the switch network. Trunk junctors for d.c. signalling are more complex because they have to provide the different signalling conditions and, in some circumstances, have to provide a through path [6].

The ferrod sensors are arranged in groups of 1024, as 64 rows of 16. This permits the scanning of 16 points in one operation by transmitting a pulse of current along the selected row [7].

Processor subsystem. The control unit consists of a pair of processors working in synchronous duplex (match-mode). The control communicates with program memory and call memory (for transient data, network map, and so on) [8]. The program is stored in a read-only memory which was specially developed by Bell for the No. 1 ESS and is called a twistor store. It consists of a coordinate-addressed array of magnetic devices (the twistors) which normally give an output when they are pulsed. It is possible to inhibit the outputs by placing a small magnet on top of the device. Information is stored in this system by placing, on top of the devices, cards which have small magnets in required positions. The program word length is 44 bits of which 7 are used as error correction bits. The call store is based on conventional magnetic core memories.

All memory data are duplicated. Figure 11.8a shows, by way of example, how a three module program store is organised to store three program blocks. [9] There are two copies of each program block, with copies in different modules. The arrangement shown allows the use of an odd number of program modules. Each module is accessed by two address buses, only one of which is enabled at any one time. The output from a module is fed to two answer buses, but again only one output is enabled at a time.

The two central processors have access to the two address buses and the two answer buses, but only one address and one answer bus are enabled for a particular processor. The other processor normally has the opposite pair of address and answer buses enabled.

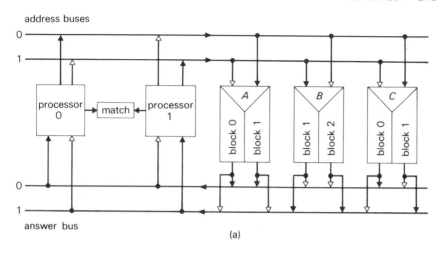

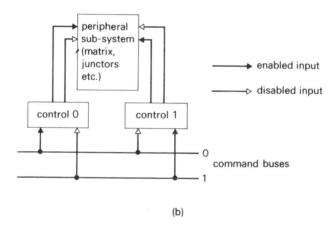

Figure 11.8 No. 1 ESS processor sub-system. (a) Example of three-program store system. (b) Control of a peripheral sub-system.

Under normal operation, each processor transmits the address of a required program instruction to its enabled address bus. As can be seen from Figure 11.8a, this address should be recognised by two modules which both place the required contents on to different answer buses back to the two processors. The addresses and the answers are continually matched to detect the occurrence of a hardware fault. In the event of such a fault, the matching circuit causes the two processors to enter a checking routine to discover the location of the hardware fault. (It may be in the processors, the buses, or a program module.)

Once the faulty sub-system is discovered, the enabling of inputs and outputs is modified to remove the faulty sub-system from service and permit the continuation of normal processing [10].

Call memory modules are 32 K of 24 bit memory and are duplicated in a similar manner to the program store.

There is a large number of scanners and distributors in a fully equipped No. 1 ESS switching centre. Each is connected to the duplicated peripheral address buses from the processor and each has a duplicated control. Each half of the control can receive its instructions from either address bus, but normally one control receives its instructions from one bus and the other control from the other bus (Fig. 11.8b). In the absence of faults, each control controls half the items in the scanner or distributor, but in the event of faults the system is automatically reconfigured to produce a working system.

1A processor. In 1976, the first, more powerful, No. 1A ESS was brought into service [11]. This uses an improved processor [12] which more than doubles the call capacity. To take advantage of this increase in capacity the No. 1A ESS switch network uses trunk switch networks of 2048 x 2048 crosspoints and up to 28 line switch networks. This gives a maximum size of around 130 000 lines.

The 1A processor is designed as a plug-in replacement for existing processors and is the control unit of the No. 1A ESS and the No. 4 ESS (described below). The original design of the No. 1 processor was conceived in the late 1950s, and by modern standards is slow. (It has an 8 μs machine cycle.) The 1A processor is based on high speed integrated circuit technology and achieves a speed increase of between a factor of 4 and 8. The program memory in this processor is the same as the call store, so it can be altered electrically.

The 1A processor has an additional set of buses which provide it with disc and magnetic tape storage. This allows a redundancy scheme in the No. 1A of less than full duplication of program and call data. Copies of the program are held on separate discs and the number of high speed program memory modules is limited to $N + 2$, rather than $2N$. In the event of a program memory module failure, a copy of the program is read from the disc memory to one of the spare program modules. The address bus port to the failed module is disabled and the spare module is instructed to recognise the addresses of the failed module.

Transient call data (including the network map) are fully duplicated, but semi-permanent data such as translation tables have only one copy on high speed memory. There are two copies on disc stores, in case they are required.

Software. No. 1 ESS is basically a function-division system [13]. The interrupt clock rate is 5 ms and some of the program tasks include:

dial pulse and digit scan	(run every 10 ms)
abandon and inter-digit pause time out	(run every 120 ms)

peripheral order (run every 25 ms)
multifrequency outpulsing (run every 25 ms)

Line scan (of 24 lines in parallel) is performed at a low priority and at an average rate of once every 100 ms. Special emergency action programs are called in when a fault is discovered by the program, or by one of the hardware detectors (such as mismatch between the data on two of the buses). While these programs are running, normal call processing is stopped and the system is unavailable. Also, since a fault can cause mutilation of stored data, it is necessary to run audit programs to check the validity of the stored data. The length of the resulting down-time depends upon the severity of the fault.

The emergency action programs are called *phases* [10]. Phase 1 is the least severe, and usually lasts only 10 seconds. It consists of rewriting constants into working memory and running several audits. It only affects calls actually in the process of being set up. Phase 2 lasts up to 45 seconds and affects all calls in the transitional state (that is between ringing and answer). The most serious emergency action is phase 3 which lasts 200 seconds and affects all calls except those in the talking state.

The present size of a typical No. 1 ESS program is 250 000 words. This is somewhat larger than the other systems described below. The main reasons for this are the relatively slow speed of the processor, the exceptionally high traffic load it is designed to take, and the range of services offered to the users and to the administration. In the early days of No. 1 ESS development, a single programmer achieved an annual software production rate of 600 debugged and documented instructions.

The development of new software tools (such as high level programming languages, use of structured programming and so on) has increased the productivity to 1000 instructions/year. There are still as many programmers working on No. 1 and No. 1A ESS as in the early days [14].

Performance. By the end of 1976 there were nearly 800 No. 1 ESS centres in service. Figure 11.9a shows how the performance has been improved with time [15]. The down-time objective (for faults affecting more than 100 lines) is 2 hours in 40 years for each switching centre. In fact, the availability of the system has also improved with time as is shown in Figure 11.9b. An analysis of the causes of system non-availability of a particular group of No. 1 ESS centres is reproduced in Table 11.1 [16]. This shows that only a few per cent of the incidents were caused by the 'classical' failure mechanism of simultaneous faults in both pairs of a duplicated systems. The majority of down-time is caused by the combined effects of human error and problems occurring during the growth of switching centres. Hardware faults in non-duplicated areas caused the next biggest number of failures.

Although software bugs caused problems in the early days of No. 1 ESS, during this period down-time due to bugs was only 4%.

274   COMPUTER CONTROLLED SWITCHING SYSTEMS

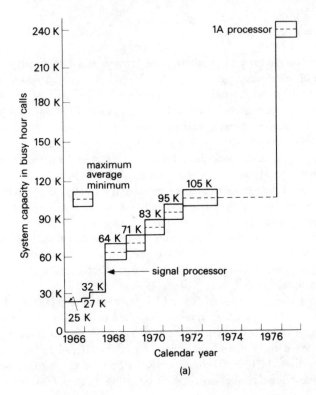

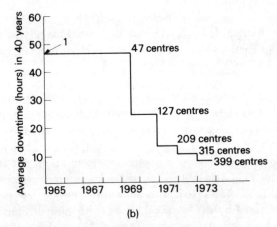

Figure 11.9   No. 1 ESS performance. (a) No. 1 ESS and No. 1A ESS capacity. (b) No. 1 ESS service performance.

Table 11.1 Causes of down time in No. 1 ESS
(Quoted in Reference [16] for the Chicago area in period 1971–3.) There were 14 switching centres in service at the end of the period, with a total capacity of over 2·5 million lines.

Causes	Restarts		Down-time	
	No.	%	Seconds	%
Total system outage requiring manual reconfiguration	2	4	7400	59
Single hardware faults	15	29	1650	13
Program bugs	5	10	530	4
Human error	11	21	1150	9
Growth	10	19	800	6
Unresolved	7	13	930	8
Traffic overload	2	4	70	1
Totals	52	100	12 530*	100

*This is equivalent to 12 hours/40 years/switching centre. If the two isolated faults are excluded, the down-time would be equivalent to 5 hours/40 years/switching centre.

The fact that the majority of the down-time is due to faults other than double faults has led to the revision of the classical reliability theory (that the system fails only if a second fault appears on half a redundant system while a first fault on the other half remains unrepaired) [17].

The new assumptions are:

— automatic recovery is successful for only a fraction of all possible single faults;
— there exists a fraction of possible faults for which automatic recovery fails, but manual recovery is possible (without repair);
— the remaining single fault (if any) causes non-recoverable system failures, and must be repaired to restore system operation.

The latter two cases account for only a very small percentage of incidents, but can produce the majority of the down-time. For these reasons Bell Laboratories have developed a number of counter-measures to attack these residual problems. They include the provision of additional monitoring information on the processor hardware and the development of portable plug-in testers, called Processor Emergency Recovery Circuits [17].

In addition, to help overcome the human error problem, Bell are moving to a system of centralised maintenance whereby a maintenance centre remotely controls a number of No. 1 ESS switching centres [18]. This concentrates the expertise of those maintaining the system in a few centres.

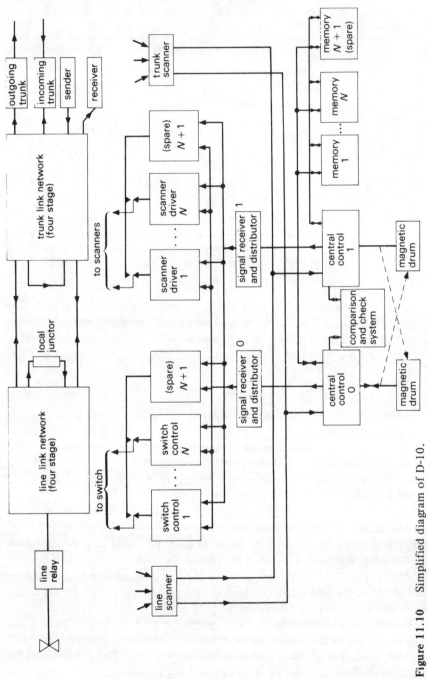

Figure 11.10 Simplified diagram of D-10.

11.3 Japanese D-10

A general view of the D-10 hardware is shown in Figure 11.10 [19]. The main features of the system are:

— complete software control;
— dual processor control;
— $N+1$ redundancy for main memories, switch and relay drivers and for scanner driving circuits.

The system was developed under the name of DEX-21 by the Electrical Communications Laboratory at the Nippon Telegraph and Telephone Company (NTT) in collaboration with the Japanese switching manufacturers.

Switch network consists of eight stages for all types of connections. The switching element is a miniature mechanically locking crossbar switch (some export versions use Remreeds) and is divided into a four-stage trunk and line link network. One line link network has a basic capacity of 2048 subscriber lines and concentrates them on to 1024 junctors. The traffic capacity of each line link network is 512 erlangs i.e. 0.5 erlang/junctor. This is the theoretical optimum discussed in Chapter 7. By additional switches, 3072, 4096, and so on, up to 8192 may be connected to one network.

The trunk link network is a four-stage network which connects 1024 junctors to 1024 trunks. The junctors from the line switch network are distributed to the trunk link network and a proportion are connected to the local junctors. The maximum size of centre consists of eight line link networks and since up to 8192 subscribers may be connected to one network, the maximum line size is 65 536. The maximum traffic of the networks is 4000 erlangs and for a normal office the processor can deal with 75 000 busy hour calls. The actual maximum size will be limited by one of these three factors. A four-wire trunk switch network version is available for trunk switching and may be incorporated with the other two networks to make trunk centres or combined local/trunk centres.

The junctors are constructed with the minimum of hardware and are based on magnetically latching relays.

Relay controllers and scanners are provided on an $N+1$ redundancy basis. When a fault is detected on one of the units, a set of change-over contacts is operated which connects in the spare module. The spare module is then instructed to respond to the addresses of the failed module.

Processor system. The original design of D-10 used a pair of processors working in synchronous duplex mode. There are two types of high-speed access memory module, plated wire electrically alterable memory for information such as programs and data, which are normally read-only and read-write core storage for the transient data. These modules contain 32 K of 32 bit words. In addition there are two high capacity (800 K) drum memories used for

- back-up copy for the main program;
- overlay programs used for administration and diagnostics;
- periodic record of call data for use in event of memory mutilation;
- trouble record, traffic measurements;
- subscribers' translation data;
- charge data.

The high-speed memories are provided on an $N+1$ basis with the drums containing back-up copies of program and transient data (recorded every 10 seconds). With $N+1$ it is not possible to use mismatch as a check for satisfactory operation and, therefore, a series of functional tests is added to the system to detect system malfunction. The simplest test is an external (triplicated) test call maker. This makes a test call at regular intervals and can detect reception of dial and other tones. This detects non-operation of systems. A variety of the tests are performed at different levels and these are described in the references [20].

The reliability of the recovery system was found to be such that working the processors in synchronous duplex mode was found to be unnecessary and the present configuration uses only worker/standby for the processors [21].

More recently, a new high-speed processor was introduced which is compatible with the original processor but four times faster [22]. These processors use only semi-conductor memory for the program and data memories. The higher computing power is required to increase the traffic capacity (of trunk centres) and provide adequate power for the provision of new services, common channel signalling and for the introduction of high-level languages.

Operational software. The main feature of the operational programs in D-10 is the extensive use of state transition diagrams as a means of design and implementation of the call processing program. An example of the form of the state transition diagram in the D-10 is shown in Figure 11.11. Points to note are:

(a) Each state box contains a diagramatic representation of the network configuration in that state.
(b) A blocking symbol is used as a branch point between states, as the new state may depend upon the availability of a required junctor circuit or network path. Also under overload conditions the new state may have to be different from that under normal conditions.
(c) A branching symbol is introduced for cases where the new state will depend upon the number dialled, subscriber classification etc.

When complicated types of connection are required such as three-party calls the number of states required to describe the system becomes very large. In order to reduce the total number a scheme of state partitioning is used whereby a trunk circuit is divided into a calling and a called part, each with its own set of

JAPANESE D-10 279

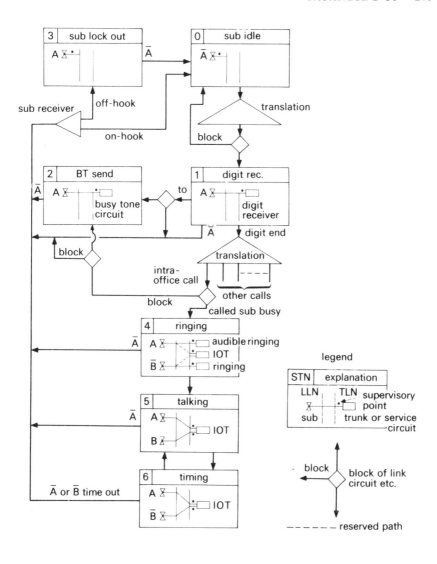

STN: state number
A: calling subscriber (receiver off-hook)
B: called subscriber (receiver on-hook)
¯: on-hook
TO: time-out
IOT: intra-office trunk (local junctor)

Figure 11.11 State transition diagram of an intra-office call (adapted from Kawashima *et al.* [23]).

states. A small number of artificial events must now be sent from one side to the other in order to co-ordinate them.

The state transition diagram provides an unambiguous and easily understood definition of the actions needed to process the telephone call. When an event occurs relating to a particular call, the central control must change the network configuration appropriately and also perform some changes on the stored information relating to the call. There may also be other functions needed, such as incrementing a traffic counter. Hence for every event leaving every state, there is a certain amount of processing required which is called a *task*. From the state transition diagram it is a straightforward procedure to define the task needed for each transition.

In the case of a local centre the number of states needed is about 110 for basic service and a further 130 for new services. The number of tasks is about

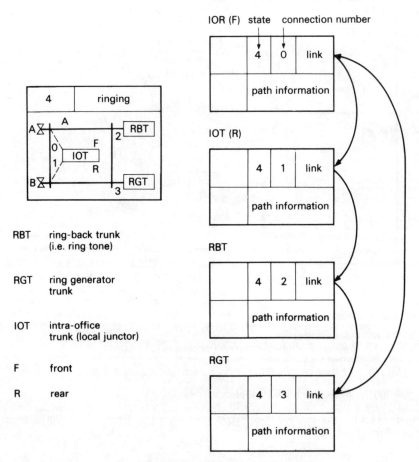

Figure 11.12 An example of the linkage of trunk memories.

300 for basic service and a further 500 for new services. This was probably the first engineering application of state transition diagrams to switching system software [23].

The D-10 system provides a direct implementation of the state transition diagram. A telephone call will always be associated with at least one junctor circuit (i.e. local junctor, register, sender, etc.) and each of these junctors is permanently allocated a number of words in the temporary memory (see Fig. 11.12). When more than one circuit is involved in a call the associated memory locations are chained together, as shown in Figure 11.12. Included in the information stored is an indication of the state of the particular call currently associated with the hardware circuit.

At regular intervals the central control scans all the junctor circuits looking for some change indicating the occurrence of an event. When an event is detected the junctor circuit number, plus an event code, are placed in a queue (Fig. 11.13).

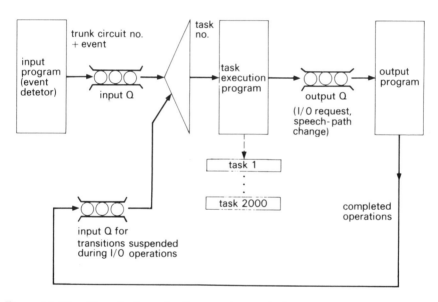

Figure 11.13   Organisation of call processing control.

This queue is serviced in an analysis program which first determines the state of the call associated with trunk circuit. Next the analysis program consults a table indexed by the current state and the new event. The entry on this table indicates the task required to perform the required transition and any other work needed. In some situations the next state will depend upon the dialled number or the terminal classification. In these cases additional analysis programs are called in order to determine the next state.

The analysis program seizes a 128 word Task Execution Memory block and copies the data from the relevant trunk memories into predetermined locations, it also inserts the next state number and task number.

Sometimes the next state cannot be reached because there is blocking due to lack of a trunk circuit or a network path. In these cases the state transition table is reaccessed with the old state and indexed by a pseudo-event code indicating blocking.

If the transition involves a change in the network connection pattern or in the trunk circuit itself, a speech path request is loaded into the output queue and the processing of task is suspended until the operation is completed. The task must also be suspended if information is needed from the drum. Once the input or output request has been completed, the suspended task is added to a second input queue and waits to be reactivated. Since the processing of a transition can take a finite time due to the input/output requests, any further events relating to that call are not processed until the existing transition is completed.

Each of the tasks consists of a combination of a much smaller number of basic operations. A considerable reduction in the program memory requirement is made by representing each of these basic operations by a single word consisting of a function field and various parameter fields. These are called task macros and a total of about 80 types are needed and, on average, each transition can be represented by 10 macros.

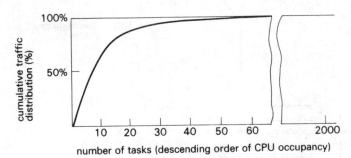

Figure 11.14 Cumulative traffic distribution versus tasks.

The price paid for this memory saving is that an interpretative program is needed to unpack the macro instruction and to decode it. This takes time and, therefore, the call handling capacity of the central control is reduced. However, it was found that 80% of the execution time of the tasks was caused by only 20 of the total number of tasks (Fig. 11.14). These tasks are, therefore, coded in normal assembly language at a small increase in memory space requirements (about 4 K) but a significant improvement in call handling capacity. If all the tasks were written in optimised assembly language there would only be a further 3% improvement in call handling at a cost of about 60 K more program memory.

The complete program size is shown in Figure 11.15.

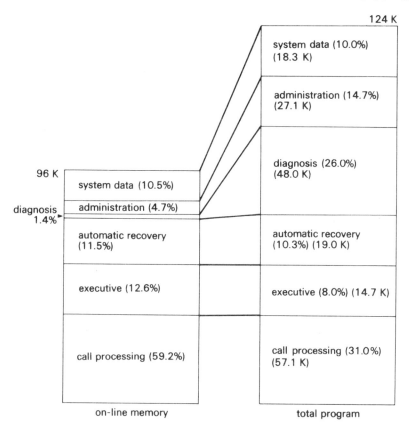

Figure 11.15 Size of first D-10 local + CENTREX + video switching program.

D-10 motherfile system. A number of different software systems are required as there are several classes of switching centre, local, trunk and local-trunk combined. Also there is the possibility of remote control sub-systems, CENTREX, video and common channel signalling. (CENTREX is a service which effectively provides PABX service from a local centre.) Also, as the centres have been introduced various hardware improvements have been made. In order to make effective management of the software for these various systems, and to minimise the problem of adding new functions to all systems, a standardised programming system called a Motherfile has been developed [24]. In this system the software needed for all the system variations is contained in a Motherfile as a number of units. A software system generation process takes those units required for a particular office configuration and function, and combines them to form the program for that office. The Motherfile contains about 460 units and the system generation process has a total of 24 parameters to specify the office type. For a simple local switching centre without extra services such as CENTREX, video or remote sub-system, the program is composed of about 360 units. A simple trunk centre system is composed of 315 units of which 257 are common with the local system.

11.4 Metaconta Systems

An alternative to dual processor working in either synchronous duplex or worker/standby mode is to use two (or more) processors in a load-sharing mode. There are three relative advantages of load sharing over the other two tech-

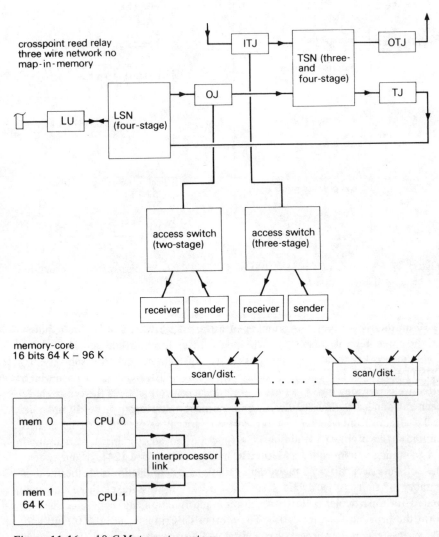

Figure 11.16 10-C Metaconta system.

niques. (a) Under normal conditions each processor takes half the traffic load, thereby improving response times. It is, however, necessary to ensure that one processor can take all the load (albeit at a degraded performance) in the event of a fault on the other processor. (b) It is possible to change the software in one processor at a time by moving the traffic from one processor to the other. (c) It is easier to detect the effect of software bugs as each processor is normally dealing with a different set of data.

The Metaconta family of switching systems produced by ITT are examples of the use of load sharing [25]. Figure 11.16 shows one member of the family, the 10-C local switching machine [26]. In this system each processor is normally looking for new calling conditions for half the time and service calls that it finds. To avoid simultaneous seizures of the same terminal or junctor, an electronic arbiter circuit is required. In this system the network map is kept in each processor and dual seizures of the same path are avoided by the processors interchanging messages via an inter-processor link. This link is also used to update each processor of the call state of the call handled by the other processor. This permits one processor to take over the traffic of the other in the event of a fault.

This load-sharing principle has been used with success in a wide variety of switching systems [27].

References

1. Hills, M. T. and Kano, S. (1976) *Programming Electronic Switching Systems.* Peter Peregrinus.
2. Feiner, A. and Hayward, W. S. (1966) No. 1 ESS switching network plan, *Bell Syst. Tech. J.*, 63, 5, pp. 2193–220.
3. Carbaugh, D. H. (1966) No. 1 ESS call processing, ibid, pp. 2483–532.
4. The original No. 1 ESS is described in a two part special issue of *Bell Syst. Tech. J.*, 63, 5, pp. 1831–2609 (Sept. 1966).
5. Special issue on Remreed switching networks for No. 1 and No. 1A ESS, *Bell Syst. Tech. J.*, 55, 5, (May–June 1976).
6. Biddulph, R. *et al.* Line, trunk, junctor and service circuits for No. 1 ESS, ibid, pp. 2321–53.
7. Freimanis, L. *et al.* No. 1 ESS scanner, signal distributor and central pulse distributor, ibid, pp. 2255–82.
8. Doblmaier, A. H. and Neville, S. M. (1969) The No. 1 ESS signal processor, *Bell Lab. Record*, 47, 4, pp. 120–4.
9. Connell, J. B. *et al.* No. 1 ESS Bus system, ibid, pp. 2021–54.
10. Downing, R. W. *et al.* No. 1 ESS maintenance plan, ibid, pp. 1961–2020.
11. Nowak, J. S. (1976) No. 1A – A new high capacity switching system, *International Switching Symposium*, paper 131–1.
12. Staehler, R. E. (1972) 1A Processor – a high-speed processor for switching applications, *International Switching Symposium*, pp. 26–35.
13. Carbaugh, D. H. No. 1 ESS call processing, ibid, pp. 2483–532.
14. Ketchledge, R. W. (1976) Programming productivity, *International Switching Symposium*, paper 232–2.
15. Fleckenstein, W. O. (1974) Bell System ESS family – present and future, *International Switching Symposium*, paper 511.
16. DeMaeyer, B. R. (1974) Introducing electronic switching equipment into the Chicago Metropolitan area, *International Switching Symposium*, paper 447.

17. Arnold, T. F. and Rohn, W. B. (1975) Minimising down-time for electronic switching systems, *Bell Lab. Record*, 53, 3, pp. 157–61.
18. Senese, D. J. and Snyder, B. E. (1974) Automated centralised maintenance techniques for electronic switching systems, *International Switching Symposium*, paper 421.
19. See special issue of *Review of the Electrical Communications Laboratories*, 19, (March 1971). Also see Kawashima, H. *et al.* Electronic Switching System Design – A Case-Study of D-10. (Provisional title, to be published by Peter Peregrinus in the IEE Telecommunications Series).
20. Nakajima, H. *et al.* (1973) Software techniques for electronic switching system dependability, *Proc. Software Engineering for Telecommunications, IEE Conference Publication 97*.
21. Honna, Y. *et al.* (1976) Evaluations of DEX reliability, design and field data, *International Switching Symposium*, paper 521–2.
22. Tamiya, T. (1977) New high-speed processor for D-10 Electronic Switching System, *Japan Telecommunications Review*, 19, 3, pp. 188–95.
23. Kawashima, H. *et al.* (1971) Functional specification of call processing by state-transition diagram, *IEEE Trans. COM-19*, 5, pp. 581–7.
24. Murata, T. *et al.* (1975) Standardised software for electronic switching systems, *Japan Telecommunications Review*, 17, pp. 93–9.
25. Kobus, S. *et al.* (1972) Central control philosophy for the Metaconta telephone switching system, *Proc. International Switching Symposium*, Boston.
26. Adelaar, H. H. (1969) The 10-C system, a stored program controlled reed switching system *IEEE Trans. COM-17*, pp. 333–9.
27. Van Os, L. *et al.* (1976) Ten Years' field experience with Metaconta systems, *International Switching Symposium*, paper 521–2.

12
Digital switching systems

12.1 Time-division switches

The cost of an analogue switch is roughly proportional to the number of crosspoints and therefore switch network design based on such switches aims to minimise the number of crosspoints. With time-division digital switching there are very different cost relationships and therefore the design of time division switch networks involves different objectives.

The principle of a digital time-division switch can be illustrated by consideration of a trunk switch interconnecting a number of p.c.m. multiplex transmission systems (Fig. 12.1a). Each p.c.m. multiplex system carries a number of channels of information (typically 24 or 32) by assigning a regular time slot to each channel. The function of the switching network is to connect pairs of channels so that information arriving at the switching centre in a particular channel on one p.c.m. multiplex system can be passed to some other channel on an outgoing p.c.m. multiplex system. To achieve this switching, two operations are required.

First, the channel must be shifted from the time slot it occupies in one system, to the one occupied in the other system. Second, it must be switched through from one system to the other. These two processes are referred to as *time switching* and *space switching*. For these to take place it is normally necessary for all transmission systems that are connected to a centre to be synchronised, so that all the time slots correspond almost exactly.

The principle of these two switching processes is illustrated in Figure 12.1a which represents just four multiplex systems with only four channels in each. The incoming channels of multiplex systems A and B are switched to the various outgoing channels of systems X and Y. The diagram shows only one half of each connection; there is a corresponding half for the other direction of transmission, that is for the outgoing channels of systems A and B, and incoming channels of systems X and Y. The time-switching process is represented by the circles, which

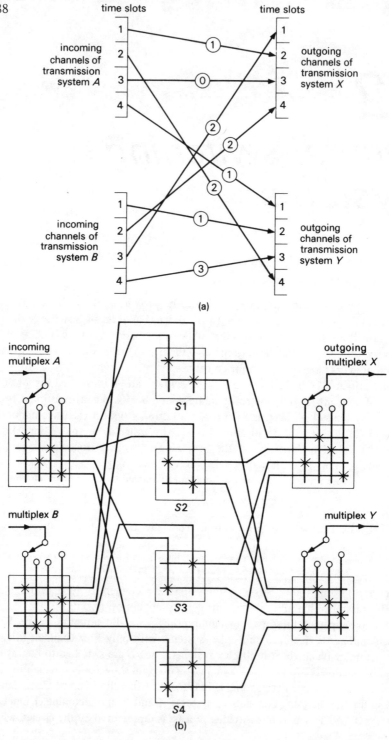

(a)

(b)

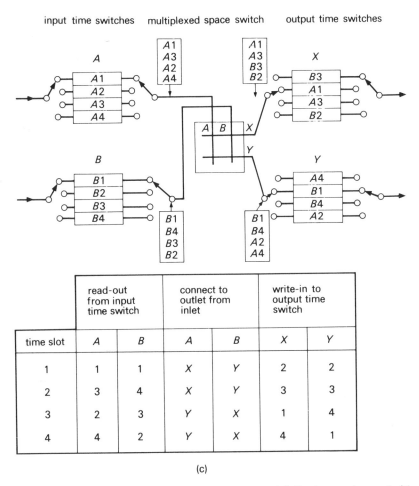

Figure 12.1 A digital TST switching network. (a) Basic requirement (the numbers indicate the number of time slots by which each channel must be delayed). (b) Simple switch network. (c) Implementation using a multiplexed space switch.

contain the number of time slots by which each channel has to be delayed. Note that, to get from time slot 2 to time slot 1, the channel is delayed by 3 time slots (since the time slots occur in rotation, so that slot 1 comes after slot 4). Where the time slots of the incoming and outgoing channels coincide, no time slot shifting is needed.

A possible switching network is shown in Figure 12.1b. The incoming multiplex systems are demultiplexed and presented to four inlets of input switches. Similarly the outgoing multiplex is created by selecting the samples from the output switches. The input and output switches are connected via a middle stage. Storage of the samples is required at the output.

This arrangement as such is not economical as it does not take advantage of the fact that the channels are already multiplexed. The input switch matrices can be built as shown in Figure 12.1c. This consists of a number of 8-bit stores, one for each incoming channel. The multiplex system writes the incoming samples into these stores in channel order.

The stored samples are read out in arbitrary order, this provides a stage of time switching as the channels are permutated. For instance, the input to the multiplex switch A is in the sequence

$$A1, A2, A3, A4$$

and the output is in the sequence

$$A1, A3, A2, A4$$

Similarly, the output sequence from multiplex switch B is

$$B1, B4, B3, B2$$

For the connection pattern in the example, it is necessary to connect channels $A1$, $A3$, $B2$, $B3$ to output multiplex switch X and the remaining channels to output multiplex switch Y. This is performed by the middle stage switch. This switch can be constructed from electronic gates and since the samples are present for only a short period (25% of the time in our simple example), the space switch can be multiplexed. In other words, Figure 12.1b shows that four space switches are required. In Figure 12.1c there is only one physical switch but it provides a different connection pattern for each time slot. In our example, in time slots 1 and 2, inlet A is connected to outlet X and outlet B is connected to outlet Y. In the other time slots the reverse connection is made. Hence the inlets and outlets of the space switch will be as follows:

$$\text{inlets} \begin{cases} A & A1 & A3 & A2 & A4 \\ B & B1 & B4 & B3 & B2 \end{cases}$$

$$\text{outlets} \begin{cases} X & A1 & A3 & B3 & B2 \\ Y & B1 & B4 & A2 & A4 \end{cases}$$

The channels from X and Y now contain the current channels for the appropriate outgoing systems but the channels are in the wrong order. The correct order is obtained by writing the channel samples into output stores in the order in which they must be transmitted. The configuration of Figure 12.1c thus achieves the requirement of Figure 12.1b.

A digital time switch using memory can be made any size without a considerable increase in cost. For instance, the example shows a 4 × 4 time switch with four channel sample stores. In addition a channel association store is required with four words of 2 bits each to specify the order in which the stored samples should be read-out (or written-in from the output time switch).

The cost of making a 32 × 32 switch is only just over a factor of 8 since this switch requires 32 8-bit stores plus a 32-word control memory of 5 bits each. A 4 × 4 space-division switch requires 16 crosspoints and a 32 × 32 requires 1024, that is a factor of 64 in cost. If there are 32 time slots, the single space switch will act as the equivalent of 32 different switches. Figure 12.2a shows an example of a trunk switch network which switches between 16 incoming and 16 outgoing 32 channel multiplex systems. The equivalent space division network, shown in Figure 12.2b would clearly be uneconomic.

An even more effective network can be built if, say, the incoming multiplex systems are taken as groups of four and treated as a single 128-channel system. The 32-channel system has a bit rate of about 2 Mbit/s, so a 128-channel system has a combined bit rate of about 8 Mbit/s. Provided the electronic circuits can operate at this speed the input time switches may be made 128 × 128 rather than 32 × 32. Also the central space switch now is the equivalent of 128 separate switches (but it has to change its connection patterns four times as often).

The higher speed network is shown in Figure 12.2c and the equivalent space division network in Figure 12.2d. With this technology it is possible to build economically a non-blocking switch.

Thus it may be seen that the design of a digital switch depends upon the cost trade-off's relating to the speed of the electronic devices used in the switch. There is no optimisation by minimising the number of crosspoints.

Types of network. The type of network described above consists of two time switches separated by a space switch. This configuration is often called a TST network. In larger systems it may be necessary to provide additional stages of space switches between the time switches. For instance, the No. 4 ESS described in the next section is a TSSSST network.

Many other configurations are possible and have been investigated from time-to-time. For instance, it is possible to have an STS network where the space switches connect channels on an incoming multiplex system to channels on an outgoing multiplex via one of a number of delay circuits. This was a popular configuration in the early days of digital switching [1] as it reduces the quantity of storage elements required. However, present day technology has made distributed memory so economic that the time-space-time arrangement is now widely adopted.

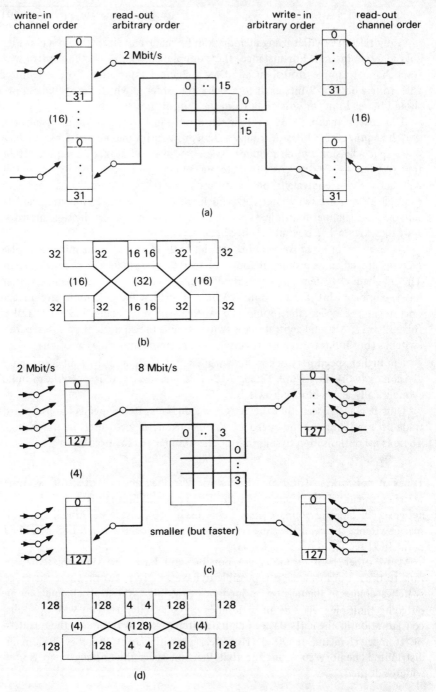

Figure 12.2 Example of digital trunk switch network. (a) Basic arrangement. (b) Equivalent space-division network. (c) Higher speed version of (a). (d) Equivalent space-division network of (c).

12.2 Relative merits of digital and analogue switching

The main advantage of digital switching is obtained if the transmission is by digital multiplex and if the digit stream is switched directly; in this case a much improved transmission performance is produced. From the transmission viewpoint, the ideal situation is where the telephone handset produces (and uses) a digital signal directly and where all the transmission and switching is performed digitally without any intermediate conversion to analogue form. In this case there is no loss or distortion introduced by the whole switching system.

A secondary advantage of digital switching is that it uses electronic rather than electro-mechanical components. This means that a system can be constructed by standard production techniques (that is automated printed circuit production and so on).

Generally, a digital switch is physically smaller than its analogue equivalent. The main problem of digital switching are in the interface cost to standard telephone instruments. Analogue to digital converters are still expensive items although the late 1970s show the development of economic single channel converters on a single chip. More troublesome is the necessity to provide the correct electrical interface to a telephone line for feeding current and ringing current. It is also necessary to protect sensitive electronic circuits from the high voltages sometimes induced on long telephone lines. These problems mean that if a digital switching system is introduced with a network of conventional telephone terminals, special (and expensive) provision has to be made for the interface. The cost of these interfaces is such that it is generally uneconomic to provide the interfaces on a per-line basis in order to permit a terminal-to-terminal digital path. Many practical systems use an analogue switch to concentrate the traffic before presenting it to an interface.

An alternative solution to the interface cost problem is to use an all-electronic telephone terminal which does not need a high operating current from a central battery or a high voltage ringing current. This solution is economic for new self-contained systems but not for existing systems.

A further problem with digital switching used with digital transmission is that of synchronisation. It is necessary for all p.c.m. multiplex systems to operate at the same frequency in a switch network consisting of a number of digital switching centres interconnected by digital transmission systems. If there is no synchronisation it means that occasionally a speech sample will be lost or repeated. In order to minimise this probability either all digital switching systems must operate from a basic clock which has a very high stability (1 part in 10^{10} typically), or each centre must have a clock which can be adjusted by some telemetry system so that the mean frequency of all centres is the same over a given period of time. There have been many studies and experiments with the objective of discovering the most appropriate method of network synchronisation. As yet a best method has not emerged.

294 DIGITAL SWITCHING SYSTEMS

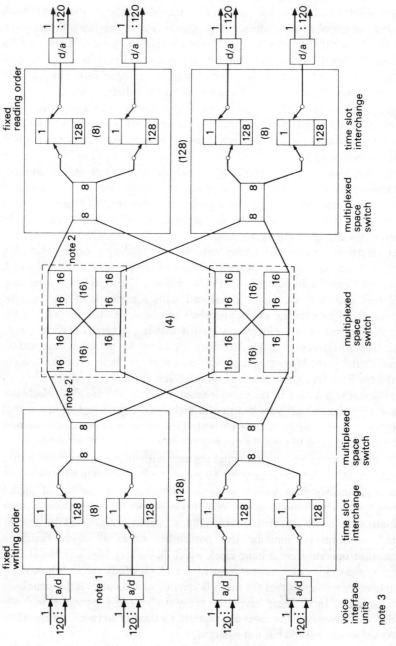

Figure 12.3 No. 4 ESS switch network. *Note 1.* The eight input analogue to digital converters are seven active with one spare. The rest of the network is completely duplicated. *Note 2.* There are two trunks from each time slot interchange to each of the four multiplexed space switches. *Note 3.* The a/d and d/a converters may be replaced by a unit which combines five 20 channels p.c.m. multiplex digital streams.

12.3 Practical application of digital telephone system – Bell No. 4 ESS

The first No. 4 ESS trunk switching system was placed into service in 1976 [2]. The No. 4 system has possibly the largest traffic capacity of all switching systems and is designed to switch up to 100 000 trunks with a traffic capacity of up to 500 000 busy hour calls [3] (equivalent to more than 40 000 erlangs). The general view of the system is shown in Figure 12.3 [4]. One hundred and twenty analogue trunks are converted to 128 channel p.c.m. multiplex digital streams by a unit called the *voice interface unit*. This digital stream is written into a memory in a fixed order. The output from this memory is read as specified by a control memory. As seen from the diagram, there are four stages of space switching before a final stage of time switching. The network is therefore TSSSST. The voice interface units are in groups of eight, of which one is a spare that can be switched into service in the event of a fault. The rest of the network is fully duplicated.

Each time slot interchange circuit switches traffic arising from effectively seven voice interface units. Since there may be up to 128 of the time slot interchange units the switch network has a maximum capacity of 107 520 terminations. Some of these terminations are required for service circuits (tones, signalling receivers, operator positions and so on).

Where the trunk transmission circuits are provided by p.c.m. multiplex it is possible to replace the analogue to digital (a/d) and digital to analogue (d/a) convertors by a unit which combines five 24-channel multiplex systems into a single digital stream. This is obviously more economic and provides the basis of providing a complete digital network in the future.

The control is by a 1A processor of the same design used on the No. 1A ESS (Ch. 11). A number of wired logic signal processors are used to relieve the main processor of the routine time-consuming tests associated with scanning and signal distribution.

References

1. Chapman, K. J., Hughes, C. J. (1968) A field trial of an experimental pulse-code modulation tandem exchange, *Post Office Elect. Eng. J.*, 61, p. 186.
2. Vaughan, H. E. and Giloth, P. K. (1976) Early No. 4 ESS field experience, *Proc. International Switching Symposium* Japan paper 241-4.
3. Vaughan, H. E. *et al.* (1976) No. 4 ESS – a fully fledged toll switching mode, *Proc. International Switching Symposium*, Japan, paper 241-1.
4. Vaughan, H. E. (1972) An introduction to No. 4 ESS, *Proc. International Switching Symposium*, Boston.

13
The future – a personal opinion

13.1 A critique of switching system design

We are now well into the second decade of stored program control of telephone switching centres. By the end of 1976 the Bell System alone had over 1000 such systems in public service and throughout the world there were possibly as many again. The original motives for introducing s.p.c. were:

(a) The increased traffic capacity that they were capable of carrying.
(b) The improved ability for functional and operational modification.

Possibly the most important differences between designs is the manner in which they achieve security with economy.

The early systems in service were all similar in structure to the Bell System No. 1 ESS and consisted of effectively a single controller constructed from suitably duplicated processors. This structure may be referred to as 'monolithic s.p.c.' [1]. The first variations were the addition of a separate signal processor in No. 1 ESS and the development of the ITT Metaconta range of load-sharing systems.

This type of system structure has clearly proved to be satisfactory in service, but possibly at a greater cost in hardware and particularly software complexity, than was originally anticipated.

Another major variant which emerged in the late 1960s was the multi-computer system such as the L. M. Ericsson AKE13 family which uses a number of duplicated processors, each pair controlling its own part of the switching network. In the United Kingdom work started separately by G.E.C. and Plessey on the development of 'true' multi-processors. In this concept a processing capability is envisaged as consisting of a number of potentially equal processors, each with access to the same memory banks and switching networks. The minimum economic size of these systems was such that for some time they were considered too expensive for installation in a single exchange. This led to concepts of 'area control' whereby a single processor system remotely controls a number of geographically separate switching centres.

A different line of development which also emerged was the structure typified by the French E-1 system. This structure consists of a number of special purpose micro-processors organised on a functional basis within a local centre. These micro-processors, in different locations, are co-ordinated by a central mini-computer system which performs non-real-time critical management tasks. This leads to a concept of hierarchical control.

The ready availability of commercial micro-processors has led to new system structures based on their use and generally called 'distributed control'. These structures typically use the micro-processors as decentralised devices for such purposes as scanning for significant events and controlling switch networks, but in general they are still effectively under the control of a central processor.

These then have been the publicly heralded and glamorous advances in s.p.c. However, there has been another range of developments which has been given much less public prominence but which, as explained below, probably shows the future direction of s.p.c. systems. These developments have been in the use of s.p.c. techniques to provide only a part of otherwise traditional electro-mechanical system structure. Examples of these structures include the G.E.C. Mk I processor systems in which a single processor provides 64 telephone registers and the L. M. Ericsson ARE system in which different processors provide register, sender and switch control functions. These systems are specially designed processors, but there are other examples based on conventional processors such as the PDP-11.

The author suggests that it is this latter development where the future direction lies, especially for the small and medium sized switching centres. In fact, it is probable that in another decade the early monolithic s.p.c. system and its direct derivatives will be regarded as a dinosaur. This animal grew so large that it was very clumsy and it took a long time for information about an event on its tail to reach its brain to be decoded. Once smaller forms of more flexible life emerged the dinosaur became extinct.

On the other hand, electro-mechanical based switching centres have developed a range of structures which have proved very secure [2]. This book has shown that there are only a few basic techniques used for the provision of secure control systems. These are either:

(a) Limit the number of terminals or the proportion of traffic dependent on a single control unit so that the loss of service in the event of a control unit fault is at an acceptable level; or
(b) Make the control unit itself very reliable by the provision of suitable levels of redundancy; or
(c) Allow a terminal (or group of terminals) or a traffic carrying section to have access to a number of control units so that in the event of one of the control units being faulty, it is still possible to process a call.

It is the last technique which is normally called 'distributed control' and is the normal method of constructing crossbar systems such as the AT and T No. 5 system. A previous paper [3] has already suggested that this type of structure is possibly the best suited to switching centre design.

In order to improve the security of such systems, a number of additional techniques are used:

(d) A control unit is designed so that even for the majority of fault conditions it can still be released by a *clear* signal.
(e) When one control unit requires the service of another, the choice of which one is made on a random basis, so that a faulty control unit will affect only a proportion of all call attempts.
(f) The separate control units do not within themselves contain any memory of the progress of individual calls, this has been termed the 'memoryless marker' system. This guarantees that individual control units are in fact dispensable.
(g) The holding time for a control unit is short and bounded so that an undue holding time may be interpreted as a fault and the control unit force-released.
(h) The alphabet of the messages transmitted between the control units is limited so that the faulty sequences can easily be detected and again force-release can be provided.
(i) Two-way signalling is provided between control units so that each can check the valid operation of the other.
(j) Even though a sub-system may be employing a degree of time-sharing, one physical system provides only a limited number of functions. This ensures that a fault within the control unit affects only those functions.

A monolithic s.p.c. system obtained its security by use of option (b) above, i.e. a single control constructed from a duplicated processor. This approach often implies that the majority of the software is required to deal with faults and subsequent reconfiguration and diagnosis. It also leaves open the possibility that unforeseen faults, or double faults, will bring the complete centre out of service. So, while it is demonstrably true that a monolithic system can provide satisfactory service, the cost in design effort and program storage is not small. Many of the inherent problems in such systems, and their derivatives, stem from the original basic assumption that it was a computer controlled system which was being designed. The use of suitably structured distributed control can avoid these costs by structuring the system in accordance with the problem, rather than a particular solution.

Interestingly, analogous concepts of structuring are now being developed in the computer software field itself, and it is being found that more economic and flexible software systems are being produced by careful choice of the overall structure.

13.2 Design principles for switching centre design

There is no single 'right' way to design a switching system but based on the above discussion it is possible to formulate some design rules to obtain a class of suitably structured systems. This approach starts from the Basic Switching System of Chapter 1, as it has the best security and no traffic limitations. It is possible to reduce the cost of this structure while keeping the level of security and traffic capacity above a design standard, by the following steps:

Step 1. Partition the control system into a number of sub-systems such that those parts needed less frequently in the progress of a call can be connected only as needed by means of access switches. The choice of the number and position of the partitions depends upon the savings obtained by providing fewer of the less-frequently used parts, less the cost of providing the access switches and any additional signalling between sub-systems.

Step 2. Replace the individual access and main switches by a number of switch networks. These networks require their own controllers and to maintain security it is necessary to decide upon objectives for the maximum number of lines and the maximum proportion of traffic carrying capacity that may go out of service as a result of a single malfunctioning switch controller. These objectives depend upon the size of the switching centre, the probability of a fault occurring and the subsequent time before it is repaired. This step now leads to the structure of Figure 13.1.

Step 3. In the structure of Figure 13.1 it is the register which is the most complex sub-system and the next stage in the design process is to consider the removal of a proportion (possibly nearly all) of its decision making components into a common control. This controller may operate on a scanning basis where the controller regularly checks each register for a change in its input conditions and performs an appropriate action, depending upon the state of that particular register. Alternatively, each register may be given a minimal decision making capability so that it can decide for itself that it needs the controller. There is a choice in the next step.

Step 4. Either technique (e) is applied so that the register access switches connect a new call at random to registers served by different controllers. Or, in the case of an interrupt driven controller, each register has access to a number of different controllers and it chooses the one required at random. Note that this requires technique (f) i.e. a memoryless controller.

Step 5. The next stage of resource sharing of the control systems is one that requires most discipline on the part of the designer. It is to consider the direct interconnection of the controllers for the line and access switches with those for

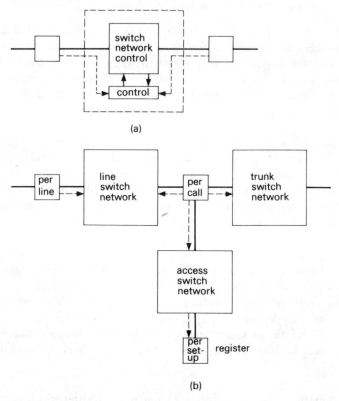

Figure 13.1 Use of common switch networks and control. (a) Building block. (b) Typical use.

the registers. The structure till now has transferred all its control information in parallel with the speech path. A direct information path between the controllers is therefore attractive, especially if they are individually realised by microprocessors. However, this can immediately lead to a system having all the problems of a monolithic s.p.c. structure unless the principles of Section 13.1 are maintained. Of these principles the most important one is thought to be the memoryless marker and the short holding time for the decision maker. Applying these principles leads to structures such as shown in Figure 13.2; in fact this is only an electronic version of the No. 5 type crossbar system [3]. The main controller now acts only as decision maker and transmitter of information. This controller may be built from a computer but it contains no state information about any call and it only processes one state transition at a time. If a mechanical switch needs to be operated, the controller is held until the operation is complete. This approach implies that no operating system is needed within the computer as this is in effect supplied by the structure of the system.

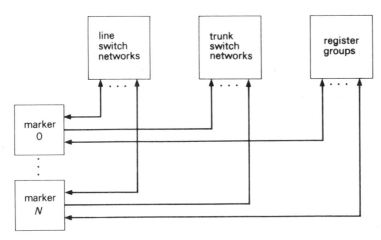

Figure 13.2 Use of memoryless markers.

It is the author's opinion that the cost trends in technology will make this the type of system structure of the future. Some exploratory design studies have been made of a system design based on these principles [4]. Time will show whether this is the approach to be adopted.

13.3 In summary

The main purpose of this book has been to show that there is an academic framework which can be used to describe switching systems. It has been shown that the main design objectives are economy coupled with high availability. A variety of techniques have been developed to satisfy these design requirements and these have been codified in the book. A switching system is a multi-processor real-time control system. Most of the design concepts were discovered and developed towards the end of the nineteenth century and early in the twentieth century.

An interesting observation is that today's computer system designers are rediscovering the principles developed by Strowger and his successors and the designs for today's large computer system are becoming increasingly similar to telephone switching centres.

Most of the switching systems described in the book have evolved around their own peculiar technology, but today's technology of large scale integrated circuits frees the designer from constraints produced by the technology. It is hoped that this book will help lay the sound theoretical background for this task.

References

1. Hills, M. T. and Lorétan R. P. (1976) The future direction of s.p.c. systems, *International Switching Symposium*, paper 412.
2. Lorétan, R. P. (1973) Security principles in electro-mechanical telephone exchanges – a comparative study, *Telecommunications Group Report No. 87*, University of Essex.
3. Hills, M. T. (1974) An alternative approach to the design of stored-program-controlled telephone exchanges, *Proc. IEE*, **121**, No. 9.
4. Lorétan, R. P. (1975) Control of telephone exchanges using multi-processor computer systems, *Telecommunications Group Report No. 134*, University of Essex.

Appendix A Optimal size of switching machine

With a number of idealised assumptions it is possible to compute the optimal size of a switching centre to minimise the total cost of transmission plus switching [1]. The assumptions are:
(1) The geographical distribution of terminals is uniform and equal to n terminals/km².
(2) The cost of local lines from a terminal to the local switching centre is a linear function of distance, that is line cost, L, is given by:

$$L = al \tag{A.1}$$

where l is the length of the line in km and a the cost in units of money per km.
(3) The cost of a switching machine plus its trunks is a linear function of the traffic that is

$$S = b_0 + b_1 z \tag{A.2}$$

where b_0 is a fixed cost per centre and b_1 is the incremental cost per erlang. z is the total traffic in erlangs.

Average length of line. If the area served by a machine is circular, the number of terminals at distances between l and $l + dl$ km from the centre is $n2\pi l\,dl$, so the total length of lines for an area with a radius of R km is

$$l_{tot} = 2\pi n \int_0^R l^2\,d = \frac{2}{3}\pi n R^3$$

The total number of terminals within the area is $N = n\pi R^2$ so the average length, $\bar{l}$, is

$$\bar{l} = \tfrac{2}{3} R$$

The total area A, is πR^2, so the average line length in terms of A is

$$\bar{l} = \frac{2}{3}\left(\frac{A}{\pi}\right)^{1/2} = 0.376\,A^{1/2}$$

304 APPENDIX A

It may be shown that, for a hexagonal or square area, the average length of line is given by

$$\bar{l} = 0.377 A^{1/2} \text{ (for hexagonal area)}$$
$$\bar{l} = 0.382 A^{1/2} \text{ (for square area)}$$

These are not very different from the circular area case.

The above analysis assumed that the actual length of line is the point-to-point distance between a terminal and the machine. However, practical routing problems make it more likely that the actual length of line will be longer than the point-to-point distance. In an urban area a typical figure of the ratio of actual length to radial length is 1·25. Hence, a more practical assumption is that the average length of local line is:

$$\bar{l} = 1.25 \times 0.377 A^{\frac{1}{2}}$$
$$\simeq 0.5 A^{\frac{1}{2}} \tag{A.3}$$

This is a useful approximation for practical purposes, provided there is a uniform density of terminals.

Multi-centre systems. If an area A is split into m equal districts with a switching centre located at the centre of each area, the average length of local line will be

$$\bar{l} = 0.5 \left(\frac{A}{m}\right)^{\frac{1}{2}}$$

Assuming that all the centres have trunks to one trunk centre at the centre of the whole area, the average length of trunks is $0.5 A^{1/2}$.

If each terminal generates a total of E erlangs/terminal (incoming plus outgoing) the total traffic originating in one district will be nEA/m erlangs.

Assuming that any one terminal has an equal probability of making a call to any other terminal within the whole area, $1/m$ of the traffic from a district will be switched within the local centre and $(1 - 1/m)$ will be switched via the trunks, that is

$$\text{Local traffic} = \left(\frac{nEA}{m}\right) \frac{1}{m} \quad \text{erlangs}$$
$$\text{Trunk traffic} = \left(\frac{nEA}{m}\right) \left(1 - \frac{1}{m}\right) \quad \text{erlangs}$$

From Equation (A.1) the cost of the local lines is $L = anA\bar{l}$ and from (A.2) the cost per switching centre is

$$S = b_0 + b_1 \left(\frac{nEA}{m}\right)$$

The total cost, C, is therefore given by

$$C = anA\,0.5 \left(\frac{A}{m}\right)^{1/2} + mb_0 + b_1 nEA \qquad (A.4)$$

(ignoring the cost of the trunk centre).

The total cost has a minimum value when $dC/dm = 0$ that is when

$$m = A \left(\frac{an}{4b_0}\right)^{2/3}$$

So the optimal area covered by one machine (under these very idealised assumptions) is

$$A_{opt} = \frac{A}{m} = \left(\frac{4b_0}{a}\right)^{2/3} n^{2/3}$$

That is the optimal area is a function of the geographical density of terminals to the power $\frac{2}{3}$. The precise optimum depends upon the ratio of the fixed cost per switching centre to cost per distance of local lines.

References

1. Boer, J. de (1967) Economic aspects of multi-exchange telephone network planning. *Philips Telecomm. Review*, **26**, p. 157. see also *CCITT Handbook National Telephone Networks for Automatic Service*, Ch. V, p. 49.

Appendix B Derivation of some basic traffic theory results

B.1 Basic assumptions

The general description of a resource sharing system consists of

(a) descriptions of the probability of arrival of new calls from a group (possibly infinite) of sources.
(b) a number, M, of servers.
(c) a probability distribution of service times of the servers.
(d) a number, Q, of waiting positions for call arrivals which find all available servers busy (Q is zero if there is no queueing allowed).
(e) in the case of queueing, an algorithm for selecting for service a new call from the queue when a server becomes free. (This may be random, first-in first-out (FIFO) and so on.)

For a system with N servers and Q waiting places, there will be certain probabilities of it being in particular states, e.g.

Probability of system having exactly k servers occupied is

$$P(k), (k = 0, \ldots M)$$

Probability of system having all N servers occupied and q calls waiting is

$$P(k), (k = M + 1, M + q \ldots M + Q).$$

For any system

$$\sum_{k=0}^{M+Q} P(k) = 1 \tag{B.1}$$

The basic traffic and queueing theory results are based on the following assumptions:

(a) Statistical independence — i.e. the statistical parameters are all independent of time so $P(k)$ is a constant independent of time.
(b) Birth and death process — this is a term which indicates that the probability of a system changing state (i.e. adding or losing a call) depends only

DERIVATION OF SOME BASIC TRAFFIC THEORY RESULTS   307

upon that state and not upon any other previous history. It also implies that the system will only change state by adding or subtracting one call.

Mathematically the birth and death process assumes that if the system is in the kth state, at any instant in time there are (at most) three possible actions:

- a new call arrives, moving the system to the $(k+1)$th state with probability λ_k;
- an existing call terminates, moving the system to the $(k-1)$th state with probability μ_k;
- no change occurs with probability $(1 - \lambda_k - \mu_k)$.

The probability of the system being in the kth state is $P(k)$ so the probability of a change from this state is $\mu_k P(k)$ to the $(k+1)$th state and $\lambda_k P(k)$ to the $(k-1)$th state.

However, the $(k-1)$th state has a probability of $\lambda_{k-1} P(k-1)$ of moving into the kth state and to the $(k+1)$th with a probability of $\mu_{k+1} P(k+1)$. The assumption of statistical equilibrium implies that the probability of any particular state is a constant and therefore the sum of probabilities for leaving a state and moving into a state must be zero. i.e.

$$-(\lambda_0 + \mu_0)P(0) + \mu_1 P(1) = 0 \quad k = 0 \quad \text{(B.2)}$$

$$-(\lambda_k + \mu_k)P(k) + \lambda_{k-1} P(k-1) + \mu_{k+1} P(k+1) = 0 \quad 0 < k < M \quad \text{(B.3)}$$

$$-\lambda_M P(M-1) + \mu_M P(M) = 0 \quad k = M \quad \text{(B.4)}$$

The main traffic equations used in switching system design may be derived from these equations by suitable choice of the λ_k and μ_k coefficients.

B.2 Infinite number of sources, blocked calls lost, no queue (Erlang-B formula)

If there is an infinite number of sources generating a total traffic of A erlangs and the average holding time is h units of time, the probability of a call arising per unit time is A/h. This is independent of the number of calls already in progress. So:

$$\lambda_k = \frac{A}{h} \quad \text{(B.5)}$$

If the holding time of a server is described by a negative exponential distribution with a mean of h units of time, the probability of the call terminating is $1/h$ per unit time independent of how long the source has already been served. So, if there are k sources connected to servers the probability that *one* of them terminates is:

$$\mu_k = \frac{k}{h} \quad \text{(B.6)}$$

Using the values of (B.5) and (B.6) in (B.2) gives $P(1) = AP(0)$. From Equation (B.3) for $k = 1$

$$P(2) = \frac{1}{2} A^2 P(0)$$

Further evaluations of (B.3) for $k = 2 \ldots N$ give

$$P(k) = \frac{A^k}{k!} P(0) \qquad (B.7)$$

Since there is no queue, the only possible states correspond to $k = 0 \ldots N$. Hence

$$\sum_{k=0}^{N} P(k) = 1$$

and therefore (from B.7)

$$P(k) = \frac{\dfrac{A^k}{k!}}{\sum_{x=0}^{M} \dfrac{A^x}{x!}} \qquad (B.8)$$

which is known as the Erlang-B equation.

Although this derivation has been made on the assumptions of a negative exponential distribution of holding time, it can be shown that the Erlang-B equation is valid for any distribution of holding times [1].

B.3 Limited sources (Engset equation)

If there is a limited number of sources, the probability of a call arising will depend upon the number of sources already being served.

When the total traffic offered from a group of N sources is A erlangs, this corresponds to a terminal occupancy of $\rho = A/N$. The proportion of the time for which a source is idle is $(1 - \rho)$ so the probability of a call arising from an idle source is ρ_m/h where

$$\rho_m = \frac{\rho}{1 - \rho} = \frac{A}{N - A}$$

For small values of occupancy $\rho_m \simeq \rho$, but as ρ becomes larger the idle time decreases. So, for the same calling rate, the probability of a call arising in the idle time will increase.

If there are k sources busy, there are $N - k$ idle. Hence the probability per unit time of *one* call arising when k sources are busy is

$$\lambda_k = (N - k) \rho_m/h \qquad (B.9)$$

DERIVATION OF SOME BASIC TRAFFIC THEORY RESULTS

Note that

$$(N-k)\rho_m = (N-k)\frac{A}{N-A}$$

so that as $N \to \infty$ then $\lambda_k \to A/h$ as in (B.5).

The probability per unit time of one call terminating when there are k busy servers is

$$\mu_k = k/h \qquad (B.10)$$

Solving the equations in (B.3) and (B.4) (in a similar manner to the Erlang-B example) with λ_k and μ_k given by (B.9) and (B.10) gives

$$P(k) = \frac{\binom{N}{k}\rho_m^k}{\sum_{x=0}^{M}\binom{N}{x}\rho_m^x} \qquad (B.11)$$

which is the Engset equation for limited sources. Note that as $N \to \infty$ then

$$\binom{N}{x}\rho_m^x \to \frac{A^x}{x!}$$

Since

$$\binom{N}{x} \to N^x \quad \text{and} \quad (A/N - A)^x \to A^x/N^x$$

Hence the Engset equation tends to the Erlang-B equations, as is expected.

B.4 Call congestion

Call congestion is the ratio of the expected number of calls which arise (in a given time) and find blocking to the expected total number in the same time. The probability of a new call arising when there are N sources of which k are busy is λ_k, given by (B.8). Hence the probability of a new call arising in a short time, δt seconds, is $(N-k)\lambda_k\delta t$. The expected total number of calls, N_T is therefore

$$N_T = \sum_{k=0}^{k=N} P(k)(N-k)\lambda_k\delta t$$

The expected number of blocked calls in the same time is

$$N_B = \sum_{k=M}^{k=N} P(k)(N-k)\lambda_k\delta t$$

So the call congestion is

$$B_C = \frac{\sum_{x=M}^{x=N} P(k)(N-k)}{\sum_{x=0}^{x=N} P(k)(N-k)} \quad (B.12)$$

For the binomial distribution where $P(k)$ is given by Equation 4.2, Equation B.12 simplifies to

$$B_C = \sum_{k=M}^{N-1} \binom{N-1}{k} \rho^k (1-\rho)^{N-1-k} \quad (B.13)$$

which, for the simple example in Figure 4.2c gives 0·40.

For the Engset distribution of Equation B.9 because blocked calls are cleared $P(k) = 0$ for $k \geq M+1$. Hence the call congestion of Equation B.12 simplifies to

$$B_C = \frac{\binom{N-1}{M} \rho_m^M}{\sum_{x=0}^{M} \binom{N-1}{x} \rho_m^x} \quad (B.14)$$

B.5 Queueing systems

Erlang-C. In a queueing system with Q waiting places Equations (B.2) and (B.3) are still valid but (B.4) is replaced by

$$-\lambda_M P(k-1) + \mu_M P(k) = 0 \quad M < k \leq M+Q \quad (B.15)$$

For infinite sources and a negative exponential holding time, the solutions to (B.2), (B.3) and (B.15) are

$$P(k) = \begin{cases} \dfrac{M^k}{k!} \eta^k P(0) & 0 \leq k \leq M-1 \\[1em] \dfrac{M^M}{M!} \eta^k P(0) & M \leq k \leq M+Q \end{cases} \quad (B.16)$$

where η is the average traffic offered to each server ($\eta = A/M$) and

$$[P(0)]^{-1} = \sum_{x=0}^{M-1} \frac{M^x}{x!} \eta^x + \frac{M^M}{M!} \frac{1-\eta^{Q+1}}{1-\eta} \eta^M \quad (B.17)$$

For an infinite queue ($Q \to \infty$) Equation (B.17) becomes

$$[P(0)]^{-1} = \sum_{x=0}^{M-1} \frac{M^x}{x!} \eta^x + \frac{M^M}{M!} \frac{\eta^M}{1-\eta} \quad (B.18)$$

In terms of A then $\eta = A/M$ so for the infinite queue the probability of k calls in the system is

$$P(k) = \begin{cases} \dfrac{A^k}{k!} P(0) & 0 \leq k \leq M \\ \dfrac{A^M}{M!} \left(\dfrac{A}{M}\right)^{k-M} P(0) & M+1 \leq k \leq \infty \end{cases} \quad (B.19)$$

where

$$P(0) = \sum_{x=0}^{M} \dfrac{A^x}{x!} - \dfrac{A^M}{M!} \left(\dfrac{A}{M-A}\right)$$

The time congestion is

$$B = \sum_{k=M}^{M+Q} P(k)$$

and for infinite sources this is equal to the call congestion.

For a queue of infinite length

$$B = \dfrac{M^M}{M!} \dfrac{\eta^M}{1-\eta} P(0) \quad (B.20)$$

This is known as the Erlang-C equation. In order to compare it with the Erlang-B equation, put $\eta = A/M$

$$B = \dfrac{\dfrac{A^M}{M!}}{\sum_{x=0}^{M} \dfrac{A^x}{x!} - \dfrac{A}{M} \sum_{x=0}^{M-1} \dfrac{A^x}{x!}} \quad (B.21)$$

The second term in the denominator makes Equation B.21 give a higher probability of blocking than that for the zero length queue Equation B.8. This occurs because in a queueing system, no calls are rejected and therefore when a call clears from a server there is now the possibility of another call waiting in the queue to take its place immediately. When there is no queue, a cleared call is not replaced until a new call arises. Hence the proportion of the time for which the system is blocked is increased when there is a queue. On the other hand (provided $A < M$) all the offered traffic is carried.

Reference
1. For example, Chapter 3 of Kosten, L. (1973) *Stochastic Theory of Service Systems*. Oxford: Pergamon.

Appendix C   Lower limit on number of crosspoints needed in multi-stage networks

Consider a general K-stage network consisting of m $(n \times n)$ matrices in each stage. The number of inlets and outlets is $N = mn$ for each stage and the total number of crosspoints is therefore

$$X = Kmn^2$$
$$= KnN \quad \text{(C.1)}$$

If the traffic per inlet is a erlangs, the average number of free outlets from the network will be $N(1-a)$. In each $(n \times n)$ matrix, the average number of free links accessible from the matrix will be $n(1-a)$. Hence taking the busy links into account, the network appears to be K stages of only $n(1-a)$ by $n(1-a)$ matrices. Thus the number of outlets accessible from a first-stage switch will, on average, be only $[n(1-a)]^K$ and this number must exceed the number of free outlets for a low probability of blocking to be obtained.

That is

$$[n(1-a)]^K > N(1-a).$$

so

$$K > \frac{\ln [N(1-a)]}{\ln [n(1-a)]} \quad \text{(C.2)}$$

The total traffic carried by the network is Na erlangs so from (C.1) and (C.2) the total number of crosspoints per erlang E, must satisfy:

$$E = \frac{X}{Na} > \frac{n}{\ln [n(1-a)]} \cdot \frac{\ln [N(1-a)]}{a} \quad \text{(C.3)}$$

Now $n/\ln n(1-a)$ has a minimum value of $e/(1-a)$ when $n(1-a) = e$. The maximum practical value of n will be N when the network becomes a single-stage $N \times N$ matrix. Hence the optimum value is:

$$n_{opt} = \begin{cases} \dfrac{e}{1-a} & \text{for} \quad a \leq 1 - \dfrac{e}{N} \\ N & a \geq 1 - \dfrac{e}{N} \end{cases} \quad \text{(C.4)}$$

So

$$K_{opt} = \frac{\ln N(1-a)}{\ln n(1-a)} \quad \text{(C.5)}$$

where n is the nearest integer to $n_{opt} (\leq N)$.

For the optimal value of n, the crosspoints/erlang are limited by:

$$E > \frac{e \ln N(1-a)}{a(1-a)} \quad \text{(C.6)}$$

This expression may be minimised with respect to a. For large N the minimum occurs when $a = 0.5$. So if the traffic per inlet is 0.5 erlang, the optimal size of switch will be:

$$n_{opt} = 2e \; (= 5.43 \ldots)$$

and

$$E > 4e \ln \tfrac{1}{2} N$$

Appendix D Simplified proof of Takagi optimum channel graph theorem

Takagi's theorms 1 and 2 state that a network with a channel graph of the form

$$X\overparen{DXD}XMXMX$$

(where X is any (or no) type of stage) is superior to a network with a channel graph of

$$X\overparen{DXD}XMXMX$$

This appendix proves the simpler case of

$$\overparen{DDD}MM \text{ being superior to } D\overparen{DD}MM$$

in order to outline the method of proof. The proof is also applied to a numerical example.

Figure D.1 shows the two channel graphs. The area between the dotted lines indicates the topological differences between the two graphs. The links are labelled (using the same notation as in Takagi's original paper).

1st to 2nd stage C
2nd to 4th stage F
4th to 5th stage E

Note that there are m C-links each feeding n F-links in Figure D.1a but in Figure D.1b there are n C-links each feeding m F-links. The two graphs still correspond to complete networks with the same number of crosspoints as may be seen from Figure D.2.

We wish to show that the blocking probability of graph D.1b, B_b, is less than that of graph in Figure D.1a, B_a. It will be shown that for any particular state of occupancies of the E links that $B_b \leqslant B_a$.

Assume that for a particular state of occupancies the states of the m E-links are given by $E_i (i = 1 \ldots m)$ where $E_i = 0$ if the link is free and 1 if it is busy in this particular state.

SIMPLIFIED PROOF OF CHANNEL GRAPH THEOREM

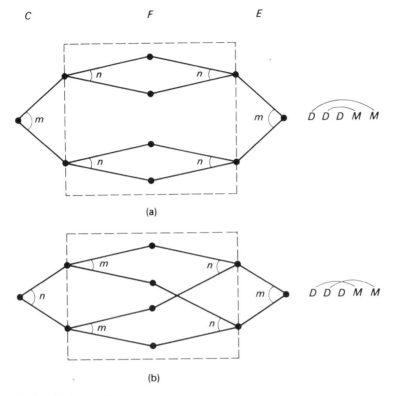

Figure D.1 Basic graphs.

The C links each have certain probabilities of being busy. Assuming that these probabilities are independent of each other, the probability of a C link being free is $P(C_0) = 1 - a$ and of it being busy is $P(C_1) = a$ where a is the traffic per input.

We may now define the blocking probabilities of the F links. The probability of there being no path via an F link connected from a C link in the ith state ($i = 0$ or 1 in this simple case) to the jth E link is defined to be $F(i,j)$.

The probability of each link being busy is a (the traffic input per line). The F links consist of two links in tandem, so the probability of both links being free is $(1 - a)^2$ and therefore the probability of there being no path via an F link is $X = 1 - (1 - a)^2$.

If an F link is connected between a C and E link which are both free, the probability of there being no path via this link will be X. If either (or both) of the C and E links are busy, the probability of there being no path via this F link is 1.

At this stage it is to introduce a numerical example to explain these concepts. This example is the simplest five-stage graph with $m = n = 2$ as shown in Figure D.3.

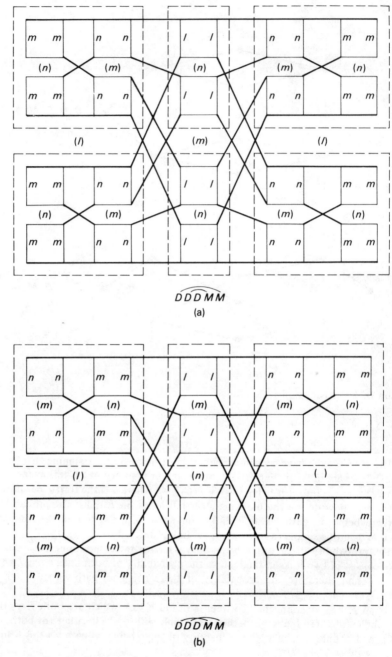

Figure D. 2 Realisation of channel graphs.

SIMPLIFIED PROOF OF CHANNEL GRAPH THEOREM

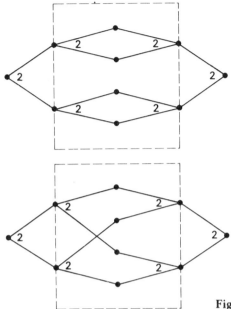

Figure D.3 Simple graphs for example.

Since there are two E links, there are four possible states, i.e.

	E_1	E_2
State 1	0	0
2	0	1
3	1	0
4	1	1

Assume (for an example) that the E links are in state 2, i.e. $E_1 = 0$ and $E_2 = 1$.

The probability that there is no path from a C link in state 0 (i.e. free) to E_1 (which is also free in this example) is $F(0,1) = X$. Since E_2 is busy, then $F(0,1) = 1$ and obviously $F(1,1) = 1$ and $F(1,2) = 1$ for all states of E.

The values of $F(i,j)$ for all four states of E are given in Table D.1.

Table D.1 Values of $F(i,j)$ corresponding to different states of the two E links

E state		F(0,1)	F(0,2)	F(1,1)	F(1,2)
E_1	E_2				
0	0	X	X	1	1
0	1	X	1	1	1
1	0	1	X	1	1
1	1	1	1	1	1

where $X = 1 - (1 - a)^2$

APPENDIX D

We are now in a position to compute the blocking probabilities for the two networks. In the first case there are n F-links in parallel from a C link to E_1. So the probability that none of these F links will offer a path through the graph is $[F(i,1)]^n$ for a particular ith state of the C link. Taking all states of the C link into account, the probability of there being no path through any of the n F-links connected to E_1 is

$$\sum_{i=0}^{1} P(C_i)[F(i, 1)]^n$$

Now there are m separate parallel paths through the graph, one via each E link. So the overall blocking probability (for a particular state of the E links) is

$$B_a = \prod_{j=1}^{m} \sum_{i=0}^{1} P(C_i)[F(i, j)]^n \qquad (D.1)$$

In the second case, each C link is connected to n F-links, one to each of the E links. The probability that there is no path via any of the F links connected to one C link is therefore

$$\prod_{j=1}^{m} F(i, j)$$

for a C link in the ith state. Taking all the states of the C link into account, the probability of no path is

$$\sum_{i=0}^{1} P(C_i) \prod_{j=1}^{m} F(i, j).$$

There are n possible paths via each of the C links, so (for the same state of the E links) the overall blocking probability for the second case is

$$B_b = \left[\sum_{i=0}^{1} P(C_i) \prod_{j=1}^{m} F(i, j) \right]^n \qquad (D.2)$$

Returning to our simple example, the expressions for B_a and B_b become

$$B_a = \{(1-p)F^2(0, 1) + pF^2(1, 1)\}\{(1-p)F^2(0, 2) + pF^2(1, 2)\}$$
$$B_b = \{(1-p)F(0, 1)F(0, 2) + pF(1, 1)F(1, 2)\}^2$$

Taking state 2 of the E links, the values are

$$B_a = \{(1-a)X^2 + a\}$$
$$B_b = \{(1-a)X + a\}^2$$

So

$$B_b - B_a = (1-a)a\left\{X^2 - 2X + \frac{1}{a}\right\}$$

$$= (1-a)a\ (X-1)^2 + \left(\frac{1}{a} - 1\right)$$

$$> 0 \text{ (for } a < 1\text{)}.$$

Hence $B_b > B_a$ and for this state of the E links the blocking probability of the graph of Figure D.1b is higher than that of D.1a.

This is a general result, it is shown in Reference [1] that for $m < n$ then

$$\sum_{i=0}^{x} P(C_i) \prod_{j=1}^{m} F(i,j) \leq \prod_{j=1}^{m} \left\{\sum_{i=0}^{x} P(C_i) F(i,j)^m \right\}^{\frac{1}{m}} \tag{D.3}$$

i.e.

$$B_b \leq B_a$$

for any state of the E links. Equality only occurs if $F(i,j) = a_i$ for all j. This implies that the state of each E link must be the same, i.e. all busy or all free. In all other cases $B_b < B_a$ so (for $m \leq n$) the graph D.1(b) will have a lower blocking probability than D.1(a). Table D.2 shows the result of a computation for our simple example with $a = \frac{1}{2}$ and binomial distribution assumed on the C links.

For $m > n$ the inequality (D.3) does not necessarily hold.

Table D.2   Values of B_a and B_b for $a = \frac{1}{2}$

E-link state		$P(E_1, E_2)$	B_a	B_b
E_1	E_2			
0	0	$\frac{1}{4}$	$\frac{625}{1024}$	$\frac{625}{1024}$
0	1	$\frac{1}{4}$	$\frac{25}{32}$	$\frac{49}{64}$
1	0	$\frac{1}{4}$	$\frac{25}{32}$	$\frac{49}{64}$
1	1	$\frac{1}{4}$	1	1
Weighted average			$\frac{3249}{4096}$	$\frac{3217}{4096}$

APPENDIX D

Specific results for networks with two mixing stages. For a five-stage network there are only two states of each of the C links, i.e. busy or free. For a C-link occupancy of a, each of

$$P(C_0) = 1 - a$$
$$= q \text{ (say)}$$
$$P(C_1) = a$$

$F(1, j) = 1$ for all j since state C_1 corresponds to a busy C link and therefore no path is possible via that F link.

$$F(0, j) = X \text{ if the } j\text{th } E \text{ link is free}$$

and

$$F(0, j) = 1 \text{ if the } j\text{th } E \text{ link is busy.}$$

Therefore if k of the m E-links are free, Equations D.1 and D.2 have the simpler form:

$$B_a = [a + q\, X^n]^k$$

and

$$B_b = [a + q\, X^k]^n$$

The total blocking probability in each case is therefore

$$B = \sum_{k=0}^{m} P(E_k)[a + q\, X^n]^k \qquad (D.4)$$

$$B = \sum_{k=0}^{m} P(E_k)[a + q\, X^k]^n \qquad (D.5)$$

for channel graph Figure D.1b.

If $P(E_k)$ is given by the binomial distribution, Equation (D.4) becomes

$$B = \sum_{k=0}^{m} \binom{m}{k} q^k\, a^{m-k} [a + q\, X^n]^k$$
$$= [a + q(q\, X^n + a)]^m$$
$$= [1 - q^2(1 - X^n)]^m$$

Equation D.5 does not have a simple reduction but obviously it will be less than D.4. It may in fact be easily evaluated with the aid of a programmable calculator.

Reference

1. Takagi, K. (1971) Optimum channel system of link system and switching network design. *Rev. Elect. Comm. Lab., Japan* **20**, pp. 962–86.

Appendix E Estimation of the traffic capacity of two-stage group selectors

The traffic capacity of a two-stage group selector may be estimated with the aid of combinational algebra by a series of techniques developed by Jacobaeus [1]. The general case is shown in Figure 4.13 (page 107) and consists of an A stage of K matrices each with n inlets to L links. These links are connected, one to each of L B-stage matrices and each B-stage matrix gives access to m_i outlets to the ith group of circuits, that is a total of $M = m_i L$ outlets in each group.

As far as a particular inlet is concerned, blocking to a required route occurs when

(a) no free link exists from the particular matrix; if the probability that any x of the L links from a particular matrix are busy is $P(x)$, the probability of there being no free link is $P(L)$;

(b) all the outlets to which a free link has access are busy.

If there are y free links from a particular A-stage matrix blocking occurs when all the my particular outlets to which these free links have access are busy. If the probability that a particular group of my of the M outlets is busy is $B(my)$ the total probability of blocking is given by

$$B = \sum_{y=0}^{L} P(L-y)B(my) \qquad (E.1)$$

An alternative, but mathematically equivalent, way of expressing the overall probability of blocking is to consider the probability, $C(x)$, that any x of the outlets are free and the probability, $Q(y)$, that y *particular* links having access to the x free outlets are busy. In which case the blocking probability is

$$B = \sum_{x=0}^{M} C(x)Q(y) \qquad (E.2)$$

The choice of equation to use will depend upon the situation and suitable choice may lead to more convenient mathematical results.

If the traffic having access to the M outlets arises from a large number of independent sources, the distribution of busy outlets may be assumed to be

TRAFFIC CAPACITY OF TWO-STAGE GROUP SELECTORS   323

given by the Erlang-B equation. Hence the probability that any x of the M outlets will be busy is

$$B(x) = E(x, A) \tag{E.3}$$

where A is the total offered traffic to the M outlets.

The probability that x *particular* outlets out of a group of M are busy may be shown to be given by:

$$B(x) = \frac{E(M, A)}{E(M - x, A)} \tag{E.4}$$

(This is known as the Palm–Jacobaeus formula.)

The probability distribution of the busy condition of the links will depend upon the type of traffic offered and the relative size of the A-stage switch. If the A-stage switch has an equal number of inlets and outlets (i.e. $n = L$) and if the inputs are from independent traffic sources, the probability that Y particular outlets are busy will be given by the binomial distribution

$$P(y) = \rho^y \tag{E.5}$$

where ρ is the average occupancy of each inlet. The probability that any y are busy will be

$$P(y) = \binom{L}{y} \rho^y (1 - \rho)^{L-y} \tag{E.6}$$

If the number of inlets exceeds the number of outlets on the A-stage switch (i.e. $n > L$) then the distributions will be described by the Engset formula (Equation B.11).

These equations may be used in three examples

Example 1. A-stage switch with $n = L$

The blocking is given in Equation E.1. The probability of Y links being busy in this case is given by (E.6) and the probability of particular my links being busy is given by (E.4) so

$$B = \sum_{y=0}^{L} \binom{L}{L-y} \rho^{L-y} (1 - \rho)^{L-y} \frac{E(M, A)}{E(my, A)} \tag{E.7}$$

There is no convenient summation for this series which must be evaluated term-by-term.

Example 2. A-stage switch with $n > L$

In this case we use the Engset distribution of Equation B.11 so

$$B = \sum_{y=0}^{L} \frac{\binom{L}{L-y} \rho_m^{L-y}}{\sum_{x=0}^{L} \binom{L}{x} \rho_m^x} \cdot \frac{E(M, A)}{E(my, A)} \tag{E.8}$$

Again, this must be evaluated term-by-term.

Example 3. A-stage switch with $n = L$ and $m = l$

The probability of blocking in this case is easier to compute if the form (e.2) is used

$$B = \sum_{x=0}^{M} E(M - x, A) \rho^x$$

In this particular case there is a summation for the series which may be found to be

$$B = \frac{E(M, A)}{E\left(M, \dfrac{A}{\rho}\right)} \tag{E.9}$$

Clearly in this case, i.e. $m = l$, there is no saving in crosspoints by using a two-stage network. However, if the switch is used as a group selector to a number of routes, each of which has one outlet per B-stage switch, then (E.9) may be seen to still apply so long as ρ is the traffic per inlet to all routes and A is the total traffic to the particular route under consideration.

Equation E.9 has a wider range of application, in Reference [2] it is explained that it is a good approximation to Example 1 and applies even if there are not equal numbers of outlets from each B switch, that is m may be a fractional number. The stated conditions under which this approximation holds is that some form of call packing is used whereby the outlets from the B-stage switch are not allocated at random, but are allocated so as to keep at least one free outlet on each B-stage switch as long as possible. In practice, however, the reference states that it is still a good approximation for pure random allocation.

For the cases where there is concentration in the A-stage switch (i.e. $n > L$) then a good approximation is

$$B = \rho^L + \frac{E(M, A)}{E\left(M, \dfrac{A}{\rho}\right)} \tag{E.10}$$

References

1. Jacobaeus, C. (1950) A study on congestion in link systems, *Ericsson Technics*, **48**.
2. Elldin, A. (1967) *Automatic Telephone Exchanges with Crossbar Switches – Switch Calculations*. L. M. Ericsson.

Index

arbiter 182–4, 187–8, 214–18
access switches 19, 60–2, 76, 219, 220–2, 243
alternative routing 119–27, 134–5
ARE System 242, 297
ARF System 239–42
audit programs 273
availability 11–12

backbone route 124
Basic Switching System 12–14, 18, 182, 299
Bernoulli distribution 86–90
Binomial distribution 86–90
blocked calls 88–90
bulk billing 137
busy hour 23–4

C 400 System 245–7
call collision 196
call division control 265–7
call records 264
CCITT No 6 Signalling System 155–9
CCITT No 7 Signalling System 159
CCS 23
central battery 27
centre of control 55, 184, 233, 243
channel graph 170–5, 314–21
charging 137, 235
circuit switching 3
Clos 169
Combination circuits 77–9
common channel signalling 155–9
common control 66–9
common switch network 58–9
compelled signalling 197
conditional selection 105–8
connector 228–30, 242–3
cost ratio 117
cross-bar
 switch 44–7
 systems 239–40
crosspoints
 definition 39
 electronic 47–8
 minimum number 166–9, 312–13
C-wire 218

D10 System 277–84
data link control procedure 35–6
destination code routing 129–30, 133
DEX 21 System 277
Dimond ring translator 212–13, 251–2
direct control 236–7
directionalisation 140
director, see register/translators
distributed control 64–7, 297–8
distributed switching 4–5
distribution frames 211–12
distribution stages 166–7

E 1 System 297
E and M signalling 147–9, 157
Electronic cross-points 47–8
EMD System 239
Engset distribution 91, 180, 308–9
entraide 249
erlang 23
Erlang-B distribution 90–2, 97, 307–8
Erlang-C distribution 92–3, 310–11
exchange code 130

far-to-near route selection 123–7
fast driver 214
feeding bridge 27–9
fereed 41–2, 269
ferrod 214–15, 269–70
final route 124
final selector 64, 235
finding schemes 57–8
folded network 161
forward transfer 144
four-wire circuits 37, 50, 127
frequency division switching 39, 52–3
full availability 97
fully provided route 124
functional sub-division 19
function division control 263–5

grade of service 24–6
grading 102–5
group selectors 64, 105–8, 233–5, 322–4
guard time 144

hierachial networks 7–10, 125–7
high usage route 119, 124

326   INDEX

hundred call seconds (CCS) 23
hunting schemes 55-6

in-band signalling 149-50
interaid 249-50
intermediate distribution frames 212
internal blocking 105, 163
international
    numbering 132
    prefix 132
    routing 127-9

Jacobaeus 323
junctors 62, 75, 222-4

Kosten 109

language digit 145
Lee 179
limited availability 102
line switch networks 72-5
line units 218-20, 241, 258-9
linked number scheme 135
load sharing 284-5
local line 27
loop disconnect signalling 29

main distribution frame 211, 212
manual hold 144
map-in-memory 260-1
map-in-network 260
marker 67, 226-31, 239-40, 242-4, 248
memoryless marker 68, 298-301
mesh networks 114
message switching 3
Metaconta System 284-5, 296
M.F. signalling receiver 225-6
M.F. signalling (subscriber) 29-31
M.F. signalling (trunk) 31-2
mismatch loss 105
mixing stages 167-9, 177-9, 320-1
Molina Traffic distribution 90
Motherfile 283-4
multiple 160

national numbering plans 131-2
network control centres 141
networks
    line switch 72-3
    non-blocking 169-70, 291
    trunk switch 72-3
network graph 170-3
network management 138-41
No. 1 ESS 267-75, 296
No. 1A ESS 268, 272
No. 4 ESS 294-5

No. 5 crossbar 242-5
non-blocking networks 169-70, 291
numbering plan areas 131, 132
number group frame 243-4

optimal channel graph 173-5
out-band signalling system 149-50
overload 98-100, 138-40
overspill traffic 120-3

packet switching 3, 35-6
Palm-Jacobaeus formula 323
p.c.m.
    signalling 151
    switching 290-1
Pentaconta system 249
Persistance testing 198
Plessey 5005 system 246-9
pre-selector 233
Poisson distribution 88-90
P-wire 218, 233-5

R1 signalling system 152-3
R2 signalling system 153-5
redundancy
    duplication 71-2
    load sharing 284-5
    triplication 72
    worker, stand-by 69-70
reed relay 40-1
registers 60
register/senders 136, 247
register-translators 83-5, 236-8
relays
    conventional 39-41, 203-11
    fereed 41-2, 269
    reed 40-1
    remreed 226-8, 270
    slugged 209-10
resonant transfer 48-50
resource sharing 18-21
restricted availability 100-5
ring around the rosie 123
ring trip 222
rotary signalling, *see* loop disconnect
roulette simulation 109-10

scanner 215-16
self steering 247
sender 75-6
separate channel signalling 155-9
serial trunking 76-7, 253-4
sequential circuits 79-80
signal exchange diagram 12-14, 181, 184, 187, 191
signalling receivers 224-6

INDEX 327

signalling
    compelled 197
    end to end 134
    link by link 134
    message switching 34-5
    telephone 27-32
    telex 32-3
    trunk 31-2
signalling paths 195
signalling system design 195-8
signalling systems
    CCITT No. 6 155-9
    CCITT No. 7 159
    common channel 155-9
    E and M 147-9, 157
    in-band 149-50
    loop disconnect 29
    m.f. (subscriber) 29-31
    m.f. (trunk) 31-2, 152-5
    outband 149-50
    p.c.m. 151
    R1 152-3
    R2 153-5
    rotary 29
signalling techniques 195-8
signals
    management 146-7
    supervisory 143-4, 147-51
    register 144-6, 151-5
simulation 108-10, 179-80
slow driver 214
space congestion 88-90
space division switching 37-48, 287, 291
s.p.c., *see* stored program control
spill over 144
star networks 114-15
Strowger system 43, 232-8
state transition diagrams 14-18, 181, 184-93, 200-4, 278-80
s.t.d., *see* state transition diagrams
step-by-step system 64-6, 67, 134-6, 232-9
store and forward 3
stored program control 83-5, 258-95
subscriber number 130
subscribers loop 27
switch control 226-31
switch networks
    line 72-5
    trunk 72-5
switching systems
    ARE 242, 247
    ARF 239-42
    C400 245-7
    centralised 7
    cross bar 239-40
    D 10 277-84
    DEX 21 277
    direct control 236-7
    E1 297
    EMD 239
    Metaconta 284-5, 296
    No 1 ESS 267-75, 296
    No 1A ESS 268, 272
    No 4 ESS 294-5
    No 5 ESS 294-5
    Pentaconta 249
    Plessey 5005 246-9
    Strowger 43, 232-8
    TXE 2 250-3
    TXE 4 253-7
    TXK 1 247-9
    TXK 2 247-9
    TXK 3 249-50
    TXK 4 249-50
synchronisation of p.c.m. systems 293

Takagi 170, 314-21
tandem switching 10, 115-17
task 16, 280-2
Telex 3
third wire control 218-22
time congestion 88-90
time division of control 20-1, 80, 258-63
time division switching 39, 48-51, 287-92
time-out 194
time true simulation 108-9
time switching 50-2, 287, 291
timing 209-11
traffic design objectives 25-6
traffic distributions
    binomial 86-90
    Erlang B 90-2, 97, 307-8
    Erlang C 92-3, 310-11
    Molina 90
    Poisson 88-90
traffic measurement 111
translators 211-3, 243, 248
transmission bridge 222-4, 235, 258-9
trombone connection 135
trunk signalling 31-2
trunk switching 7-10, 124-9
trunk switch networks 72-5
twister store 270
two motion selectors 43-5
TXE 2 250-3
TXK 1, TXK 2 247-9
TXK 3, TXK 4 249-50

uniselector 41-4, 233
    motor 238-9

variance to mean ratio 121-3

waiting time 93
watchdog timer 93
Wilkinson 122
wiper switching 42, 245, 247